Personnel

AWS Committee on Definitions and Symbols

R. L. Holdren, Chairman	Edison Welding Institute
A. J. Kathrens, 1st Vice Chairman	Canadian Welding Bureau
J. E. Greer, 2nd Vice Chairman	Morraine Valley Community College
C. B. Pollock, Secretary	American Welding Society
L. J. Barley	ITW Welding Products Company
H. B. Cary	Consultant
J. P. Christein	Newport News Shipbuilding
*G. B. Coates	General Dynamics Armament Systems
C. K. Ford	Hobart Institute
*K. W. Fordyce	Elliott Company
W. L. Green	Ohio State University
B. B. Grimmett	Ashland Chemical Company
M. J. Grycko, Jr.	Packer Engineering
J. G. Guenther	Dean Lally L P
E. A. Harwart	Consultant
*M. J. Houle	Welding Engineering Services
R. D. McGuire	National Board of Boiler and Pressure Vessel Inspectors
*D. H. Orts	Consultant
L. J. Siy	Compositools, Incorporated
J. J. Stanczak	Steel Detailers and Designers
J. J. Vagi	J. J. Vagi Consultant

AWS Subcommittee on Symbols

A. J. Kathrens, Chairman	Canadian Welding Bureau
R. D. McGuire, 1st Vice Chairman	National Board of Boiler & Pressure Vessel Inspectors
C. B. Pollock, Secretary	American Welding Society
*L. J. Barley	ITW Welding Products Company
J. P. Christein	Newport News Shipbuilding
*G. B. Coates	General Dynamics Armament Systems
C. K. Ford	Hobart Institute
W. L. Green	Ohio State University
J. G. Guenther	Dean Lally L P
E. A. Harwart	Consultant
L. J. Siy	Compositools, Incorporated
J. J. Stanczak	Steel Detailers & Designers

*Advisor

Foreword

(This Foreword is not a part of ANSI/AWS A2.4-98, *Standard Symbols for Welding, Brazing, and Nondestructive Examination*, but is included for information purposes only).

Welding cannot take its proper place as a fabricating tool unless means are provided for conveying the information from the designer to the welding personnel. Statements such as "to be welded throughout" or "to be completely welded," in effect, transfer the design responsibility from the designer to the welder, who cannot be expected to know design requirements.

These symbols provide the means for placing welding, brazing, and examination information on drawings. The system for symbolic representation of welds on engineering drawings used in this standard is consistent with the "third angle" method of projection. This is the method predominately used in the United States. In practice, many companies will need only a few of the symbols and, if they desire, can select only the parts of the system that fit their needs.

In the past, the use of the words, "far side" and "near side" in the interpretation of welding symbols has led to confusion because when joints are shown in section, all welds are equally distant from the reader and the words "near" and "far" are meaningless. In the present system, the joint is the basis of reference. Any welded joint indicated by a symbol will always have an "arrow side" and an "other side." Accordingly, the terms *arrow side, other side*, and *both sides* are used herein to locate the weld with respect to the joint.

The tail of the symbol is used for designating the welding and cutting processes, as well as the welding specifications, procedures, or the supplementary information to be used in making the weld. When only the size and type of weld are specified, the information necessary for making that weld is limited. The process, identification of filler metal that is to be used, whether peening, root gouging, or other operations are required, and other pertinent data, should be known. The notation to be placed in the tail of the symbol indicating these data will usually be established by each user.

Symbols in this publication are intended to be used to facilitate communications among designer, shop, and fabrication personnel. The usual limitations included in specifications and codes are beyond the scope of this standard.

Illustrations included with the text are intended to show how correct applications of symbols may be used to convey welding or examination information and are not intended to represent recommended welding or design practice.

Part B, Brazing Symbols, uses the same symbols for brazing that are used for welding.

Part C, Nondestructive Examination Symbols, establishes symbols to be used on drawings to specify nondestructive examination for determining the soundness of materials. The nondestructive examination symbols included in the standard represent nondestructive examination methods as discussed in the latest edition of AWS publication B1.10, *Guide for the Nondestructive Inspection of Welds*. Definitions and details for use of the various nondestructive examination methods are found in AWS B1.10.

AWS A2.4 came into existence in 1976 as the result of combining and superseding two earlier documents A2.0, *Standard Welding Symbols*, and A2.2, *Nondestructive Testing Symbols*. Both of the earlier documents had their origins in work done jointly by the American Welding Society and ASA Sectional Committee Y32. A2.0 was first published in 1947 and revised in 1958 and 1968; A2.2 first appeared in 1958 and was revised in 1969.

AWS A2.4-76, *Symbols for Welding and Nondestructive Testing*, was the first version of the combined documents and was prepared by the AWS Committee on Definitions and Symbols. It was revised in 1979 as A2.4-79, *Symbols for Welding and Nondestructive Testing, Including Brazing* and revised again in 1986 with the title, *Standard Symbols for Welding, Brazing, and Nondestructive Examination*. ANSI/AWS A2.4-98 is the second revision of the 1986 document and has the same title.

Official interpretations of any of the technical requirements of this standard may be obtained by sending a request, in writing, to the Managing Director, Technical Services, American Welding Society. A formal reply will be issued after it has been reviewed by the appropriate personnel following established procedures. Users of this standard are invited to suggest additional symbols or revisions for consideration by the committee. These suggestions should be addressed to the Secretary, Committee on Definitions and Symbols, American Welding Society, 550 N.W. LeJeune Road, Miami, Florida 33126.

Key Words — Weld symbols, welding symbols, brazing symbols, nondestructive examination symbols

ANSI/AWS A2.4-98
An American National Standard

Approved by
American National Standards Institute
November 6, 1997

Standard Symbols for Welding, Brazing, and Nondestructive Examination

Supersedes ANSI/AWS A2.4-93

Prepared by
AWS Committee on Definitions and Symbols

Under the Direction of
AWS Technical Activities Committee

Approved by
AWS Board of Directors

Abstract

This standard establishes a method of specifying certain welding, brazing, and nondestructive examination information by means of symbols. Detailed information and examples are provided for the construction and interpretation of these symbols. This system provides a means of specifying welding or brazing operations and nondestructive examination, as well as the examination method, frequency, and extent.

Reproduced By GLOBAL ENGINEERING DOCUMENTS
With The Permission OF AWS
Under Royalty Agreement

American Welding Society
550 N.W. LeJeune Road, Miami, Florida 33126

Statement on Use of AWS Standards

All standards (codes, specifications, recommended practices, methods, classifications, and guides) of the American Welding Society are voluntary consensus standards that have been developed in accordance with the rules of the American National Standards Institute. When AWS standards are either incorporated in, or made part of, documents that are included in federal or state laws and regulations, or the regulations of other governmental bodies, their provisions carry the full legal authority of the statute. In such cases, any changes in those AWS standards must be approved by the governmental body having statutory jurisdiction before they can become a part of those laws and regulations. In all cases, these standards carry the full legal authority of the contract or other document that invokes the AWS standards. Where this contractual relationship exists, changes in or deviations from requirements of an AWS standard must be by agreement between the contracting parties.

International Standard Book Number: 0-87171-524-4

American Welding Society, 550 N.W. LeJeune Road, Miami, FL 33126

© 1998 by American Welding Society. All rights reserved
Printed in the United States of America

Note: The primary purpose of AWS is to serve and benefit its members. To this end, AWS provides a forum for the exchange, consideration, and discussion of ideas and proposals that are relevant to the welding industry and the consensus of which forms the basis for these standards. By providing such a forum, AWS does not assume any duties to which a user of these standards may be required to adhere. By publishing this standard, the American Welding Society does not insure anyone using the information it contains against any liability arising from that use. Publication of a standard by the American Welding Society does not carry with it any right to make, use, or sell any patented items. Users of the information in this standard should make an independent, substantiating investigation of the validity of that information for their particular use and the patent status of any item referred to herein.

With regard to technical inquiries made concerning AWS standards, oral opinions on AWS standards may be rendered. However, such opinions represent only the personal opinions of the particular individuals giving them. These individuals do not speak on behalf of AWS, nor do these oral opinions constitute official or unofficial opinions or interpretations of AWS. In addition, oral opinions are informal and should not be used as a substitute for an official interpretation.

This standard is subject to revision at any time by the AWS Committee on Definitions and Symbols. It must be reviewed every five years and if not revised, it must be either reapproved or withdrawn. Comments (recommendations, additions, or deletions) and any pertinent data that may be of use in improving this standard are requested and should be addressed to AWS Headquarters. Such comments will receive careful consideration by the AWS Committee on Definitions and Symbols and the author of the comments will be informed of the Committee's response to the comments. Guests are invited to attend all meetings of the AWS Committee on Definitions and Symbols to express their comments verbally. Procedures for appeal of an adverse decision concerning all such comments are provided in the Rules of Operation of the Technical Activities Committee. A copy of these Rules can be obtained from the American Welding Society, 550 N.W. LeJeune Road, Miami, FL 33126.

Photocopy Rights

Authorization to photocopy items for internal, personal, or educational classroom use only, or the internal, personal, or educational classroom use only of specific clients, is granted by the American Welding Society (AWS) provided that the appropriate fee is paid to the Copyright Clearance Center, 222 Rosewood Drive, Danvers, MA 01923, Tel: 508-750-8400; online: http://www.copyright.com

Table of Contents

Page No.

Personnel iii
Foreword iv
List of Tables viii
List of Figures viii

Part A—Welding Symbols 1

1. Basic Symbols 1
   1.1 Distinction Between Weld Symbol and Welding Symbol 1
   1.2 Weld Symbols 1
   1.3 Welding Symbols 1
   1.4 Supplementary Symbols 1
   1.5 Placement of Welding Symbol 1
   1.6 Illustrations 1

2. Basic Types of Joints 4

3. General Provisions 4
   3.1 Location Significance of Arrow 4
   3.2 Location of Weld with Respect to Joint 4
   3.3 Orientation of Specific Weld Symbols 5
   3.4 Break in Arrow 5
   3.5 Combined Weld Symbols 5
   3.6 Multiple Arrow Lines 6
   3.7 Multiple Reference Lines 6
   3.8 Field Weld Symbol 6
   3.9 Extent of Welding Denoted by Symbols 7
   3.10 Weld-All-Around Symbol 7
   3.11 Tail of the Welding Symbol 7
   3.12 Contours Obtained by Welding 8
   3.13 Finishing of Welds 8
   3.14 Melt-Through Symbol 8
   3.15 Melt-Through with Edge Welds 8
   3.16 Method of Drawing Symbols 9
   3.17 U.S. Customary and Metric Units 9
   3.18 Weld Dimension Tolerance 9
   3.19 Changes in Joint Geometry During Welding 9

4. Groove Welds 21
   4.1 General 21
   4.2 Depth of Bevel and Groove Weld Size 22
   4.3 Groove Dimensions 23
   4.4 Length of Groove Welds 24
   4.5 Intermittent Groove Welds 24
   4.6 Contours and Finishing of Groove Welds 25
   4.7 Back and Backing Welds 25
   4.8 Joint with Backing 26
   4.9 Joint with Spacer 26

4.10 Consumable Inserts 27
  4.11 Groove Welds with Backgouging 27
  4.12 Seal Welds 27
  4.13 Skewed Joints 27

5. Fillet Welds 50
  5.1 General 50
  5.2 Size of Fillet Welds 50
  5.3 Length of Fillet Welds 50
  5.4 Intermittent Fillet Welds 51
  5.5 Fillet Welds in Holes and Slots 51
  5.6 Contours and Finishing of Fillet Welds 51
  5.7 Skewed Joints 51

6. Plug Welds 56
  6.1 General 56
  6.2 Plug Weld Size 56
  6.3 Angle of Countersink 56
  6.4 Depth of Filling 56
  6.5 Spacing of Plug Welds 56
  6.6 Number of Plug Welds 56
  6.7 Contours and Finishing of Plug Welds 57
  6.8 Joints Involving Three or More Members 57

7. Slot Welds 61
  7.1 General 61
  7.2 Width of Slot Welds 61
  7.3 Length of Slot Welds 61
  7.4 Angle of Countersink 61
  7.5 Depth of Filling 61
  7.6 Spacing of Slot Welds 61
  7.7 Number of Slot Welds 62
  7.8 Location and Orientation of Slot Welds 62
  7.9 Contours and Finishing of Slot Welds 62

8. Spot Welds 65
  8.1 General 65
  8.2 Size or Strength of Spot Welds 65
  8.3 Spacing of Spot Welds 65
  8.4 Number of Spot Welds 66
  8.5 Extent of Spot Welding 66
  8.6 Contours and Finishing of Spot Welds 66
  8.7 Multiple-Member Spot Welds 66

9. Seam Welds 72
  9.1 General 72
  9.2 Size and Strength of Seam Welds 72
  9.3 Length of Seam Welds 72
  9.4 Dimensions of Intermittent Seam Welds 73
  9.5 Number of Seam Welds 73
  9.6 Orientation of Seam Welds 73
  9.7 Contours and Finishing of Seam Welds 73
  9.8 Multiple-Member Seam Welds 73

10. Edge Welds 78
  10.1 General 78
  10.2 Edge Weld Size 78

	10.3 Single- and Double-Edge Welds	78
	10.4 Edge Welds Requiring Complete Joint Penetration	78
	10.5 Edge Welds on Joints with More Than Two Members	78
	10.6 Length of Edge Welds	78
	10.7 Intermittent Edge Welds	78
11.	Stud Welds	82
	11.1 Side Significance	82
	11.2 Stud Size	82
	11.3 Spacing of Stud Welds	82
	11.4 Number of Stud Welds	82
	11.5 Dimension Location	82
	11.6 Location of First and Last Stud Welds	82
12.	Surfacing Welds	82
	12.1 Use of Surfacing Weld Symbol	82
	12.2 Size (Thickness) of Surfacing Welds	82
	12.3 Extent, Location, and Orientation of Surfacing Welds	83
	12.4 Surfacing a Previous Weld	83
	12.5 Surfacing to Adjust Dimensions	83

Part B—Brazing Symbols 83

13.	Brazed Joints	83

Part C—Nondestructive Examination Symbols 89

14.	Elements of the Nondestructive Examination Symbol	89
	14.1 Examination Method Letter Designations	89
	14.2 Supplementary Symbols	89
	14.3 Standard Location of Elements of a Nondestructive Examination Symbol	89
15.	General Provisions	89
	15.1 Location Significance of Arrow	89
	15.2 Location of Letter Designations	90
	15.3 U.S. Customary and Metric Units	90
16.	Supplementary Symbols	91
	16.1 Examine-All-Around	91
	16.2 Field Examinations	91
	16.3 Radiation Direction	91
17.	Specifications, Codes, and References	91
18.	Extent, Location, and Orientation of Nondestructive Examination	91
	18.1 Specifying Length of Section to be Examined	91
	18.2 Number of Examinations	92
	18.3 Examination of Areas	92

Annex A—Design of Standard Symbols (Inches) 97
Annex AM—Design of Standard Symbols (Millimeters) 100
Annex B—Commentary on AWS A2.4-98 103
Welding Symbol Chart 106
Definitions and Symbols Document List 109

List of Tables

Table		Page No.
1	Letter Designations of Welding and Allied Processes and Their Variations	93
2	Alphabetical Cross Reference to Table 1 by Process	94
3	Alphabetical Cross Reference to Table 1 by Letter Designation	95
4	Suffixes for Optional Use in Applying Welding and Allied Processes	96
5	Obsolete or Seldom Used Processes	96
6	Joint Type Designators	96

List of Figures

Figure		Page No.
1	Weld Symbols	2
2	Standard Location of Elements of a Welding Symbol	3
3	Supplementary Symbols	3
4	Basic Joints	10
5	Applications of Arrow and Other Side Convention	11
6	Applications of Break in Arrow of Welding Symbol	12
7	Combinations of Weld Symbols	13
8	Specification of Location and Extent of Fillet Welds	14
9	Specification of Extent of Welding	16
10	Applications of "Typical" Welding Symbols	19
11	Applications of Melt-Through Symbol	20
12	Specification of Groove Weld Size Depth of Bevel Not Specified	28
13	Application of Dimensions to Groove Weld Symbol	29
14	Groove Weld Size "(E)" Related to Depth of Bevel "S"	29
15	Specification of Groove Weld Size and Depth of Bevel	31
16	Specification of Groove Weld Size Only	32
17	Combined Groove and Fillet Welds	33
18	Complete Joint Penetration with Joint Geometry Optional	34
19	Partial Joint Penetration with Joint Geometry Optional	35
20	Applications of Flare-Bevel and Flare-V-Groove Weld Symbols	36
21	Specification of Root Opening of Groove Welds	38
22	Specification of Groove Angle of Groove Welds	39
23	Specification of Length of Groove Welds	40
24	Specification of Extent of Welding for Groove Welds	41
25	Applications of Intermittent Welds	42
26	Applications of Flush and Convex Contour Symbols	44
27	Applications of Back or Backing Weld Symbol	45
28	Joints with Backing or Spacers	46
29	Application of the Consumable Insert Symbol	47
30	Groove Welds with Backgouging	48
31	Skewed Joint	49

32	Specification of Size and Length of Fillet Welds	52
33	Applications of Intermittent Fillet Weld Symbols	53
34	Applications of Fillet Weld Symbol	55
35	Applications of Plug Weld Symbol	58
36	Applications of Information to Plug Weld Symbols	59
37	Applications of Slot Weld Symbol	63
38	Applications of Information to Slot Weld Symbols	64
39	Applications of Spot Weld Symbol	67
40	Applications of Information to Spot Weld Symbol	68
41	Applications of Projection Weld Symbol	70
42	Multiple Member Spot Weld	71
43	Applications of Seam Weld Symbol	74
44	Applications of Information to Seam Weld Symbol	75
45	Multiple Member Seam Weld	77
46	Applications of Edge Weld Symbols	79
47	Applications of Stud Weld Symbol	84
48	Applications of Surfacing Weld Symbol	85
49	Applications of Brazing Symbols	86
50	Standard Location of Elements	89

Standard Symbols for Welding, Brazing, and Nondestructive Examination

Part A
Welding Symbols

1. Basic Symbols

1.1 Distinction Between Weld Symbol and Welding Symbol. This standard makes a distinction between the terms *weld symbol* and *welding symbol*. The weld symbol indicates the type of weld and, when used, is a part of the welding symbol.

1.2 Weld Symbols. Weld symbols shall be as shown in Figure 1. The symbols shall be drawn "on" the reference line (for illustrative purposes shown dashed).

1.3 Welding Symbols. The welding symbol consists of several elements (see Figure 2). Only the reference line and arrow are required elements. Additional elements may be included to convey specific welding information. Alternatively, welding information may be conveyed by other means such as by drawing notes or details, specifications, standards, codes, or other drawings which eliminates the need to include the corresponding elements in the welding symbol. All elements, when used, shall have specific locations within the welding symbol as shown in Figure 2. Mandatory requirements regarding each element in a welding symbol refer to the location of the element and should not be interpreted as a necessity to include the element in every welding symbol.

1.4 Supplementary Symbols. Supplementary symbols to be used in connection with welding symbols shall be as shown in Figure 3.

1.5 Placement of Welding Symbol. The arrow of the welding symbol shall point to a line on the drawing which conclusively identifies the proposed joint. It is recommended that the arrow point to a solid line (object line, visible line); however, the arrow may point to a dashed line (invisible, hidden line).

1.6 Illustrations. Examples given, including dimensions, are illustrative only and are intended to demonstrate the proper application of principles. They are not intended to represent design practices, or to replace code or specification requirements.

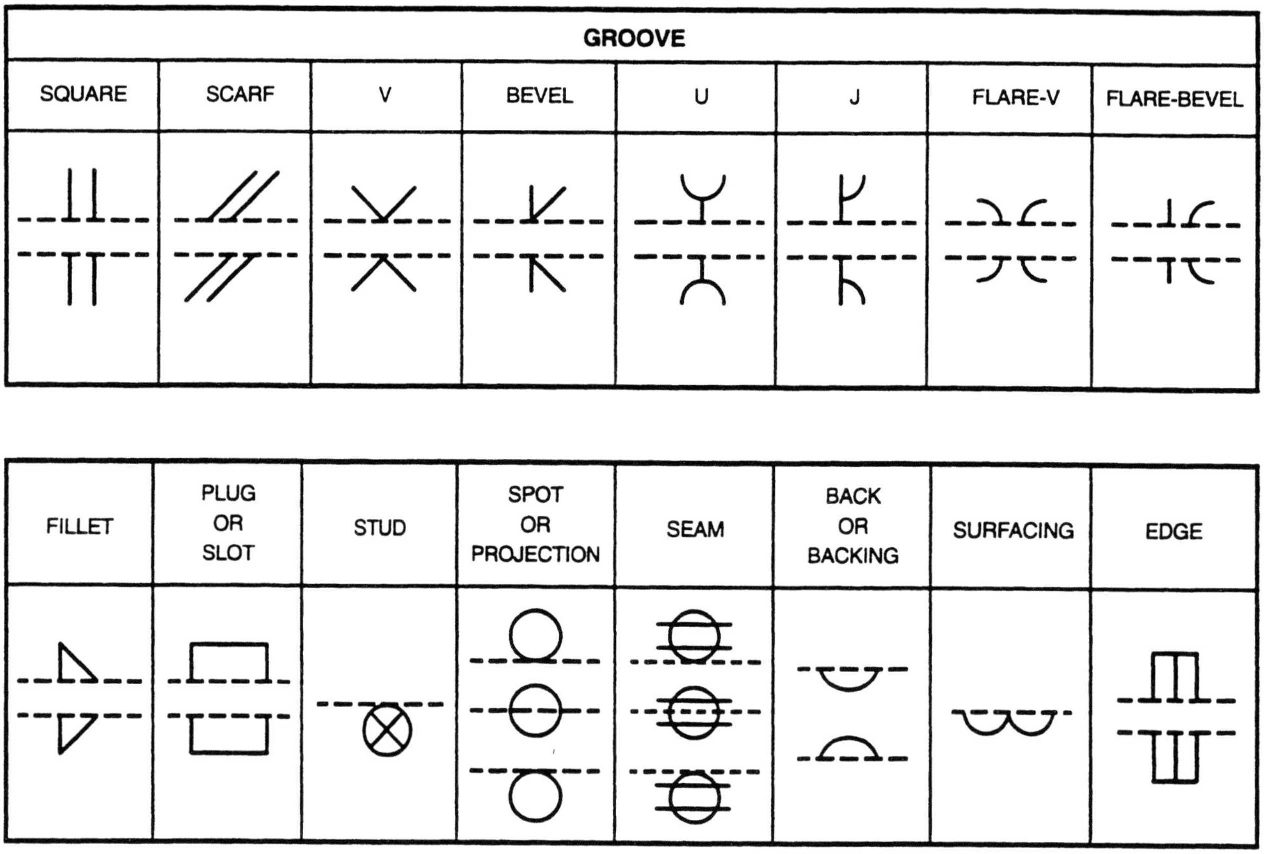

GROOVE							
SQUARE	SCARF	V	BEVEL	U	J	FLARE-V	FLARE-BEVEL

FILLET	PLUG OR SLOT	STUD	SPOT OR PROJECTION	SEAM	BACK OR BACKING	SURFACING	EDGE

NOTE: THE REFERENCE LINE IS SHOWN DASHED FOR ILLUSTRATIVE PURPOSES.

Figure 1—Weld Symbols

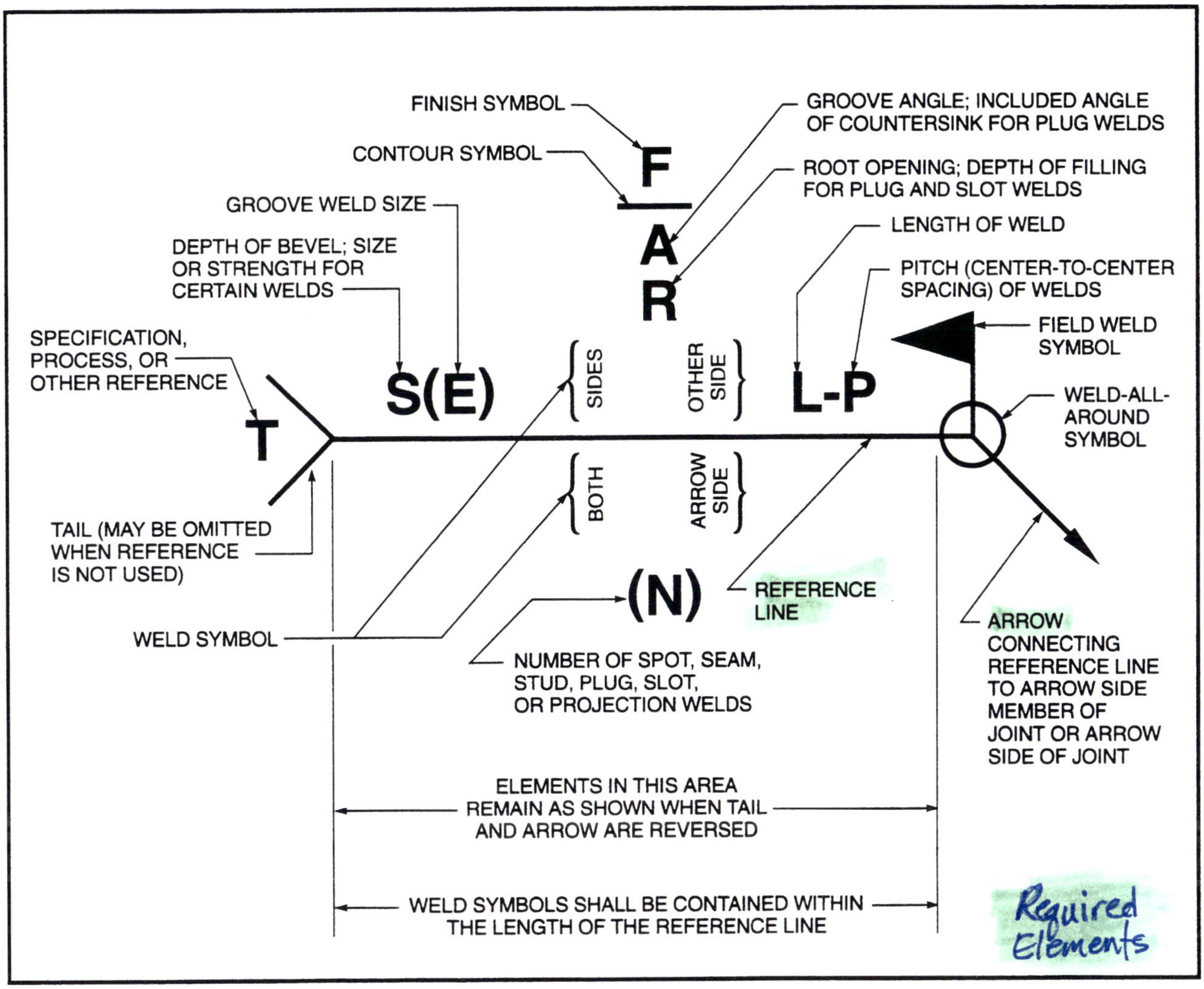

Figure 2—Standard Location of Elements of a Welding Symbol

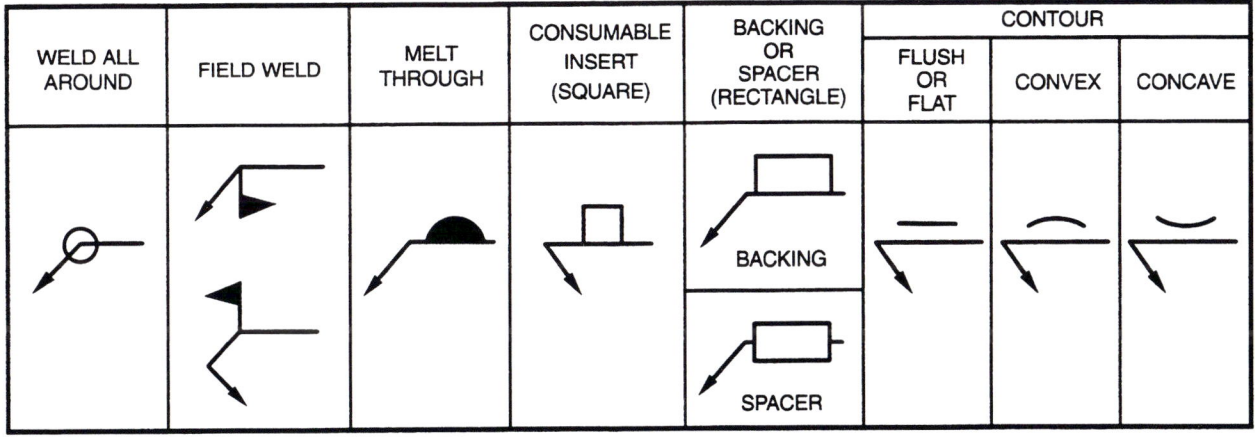

Figure 3—Supplementary Symbols

2. Basic Types of Joints

The basic types of joints are shown in Figure 4.

3. General Provisions

3.1 Location Significance of Arrow. Information applicable to the arrow side of a joint shall be placed below the reference line. Information applicable to the other side of a joint shall be placed above the reference line.

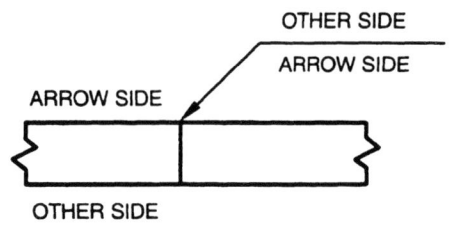

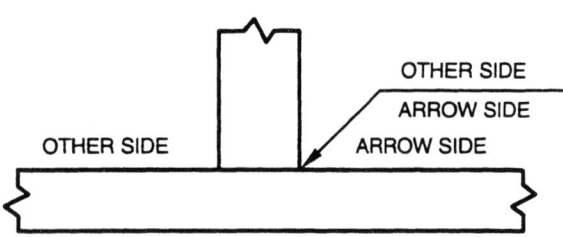

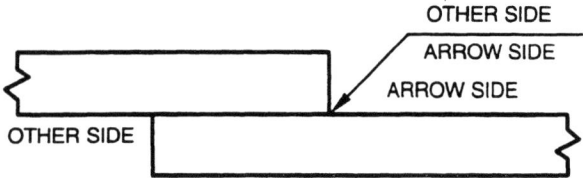

3.1.1 Fillet, Groove, and Edge Weld Symbols. For these symbols, the arrow shall connect the welding symbol reference line to one side of the joint, and this side shall be considered the arrow side of the joint. The side opposite the arrow side of the joint shall be considered the other side of the joint (see Figure 5).

3.1.2 Plug, Slot, Spot, Projection, and Seam Weld Symbols. For these symbols, the arrow shall connect the welding symbol reference line to the outer surface of one of the joint members at the centerline of the desired weld. The member toward which the arrow points shall be considered the *arrow side member*. The other joint member shall be considered the *other side member* (see Figures cited in sections 6 to 9 inclusive).

3.1.3 Symbols with No Side Significance. Some weld symbols have no arrow-side or other-side significance, although supplementary symbols used in conjunction with them may have such significance (see 8.1.2, 8.1.4, and Tables 1 and 2).

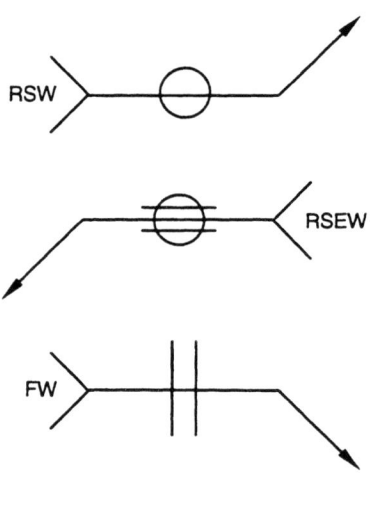

3.2 Location of Weld with Respect to Joint

3.2.1 Arrow Side. Welds on the arrow side of the joint shall be specified by placing the weld symbol below the reference line (see 3.1.1).

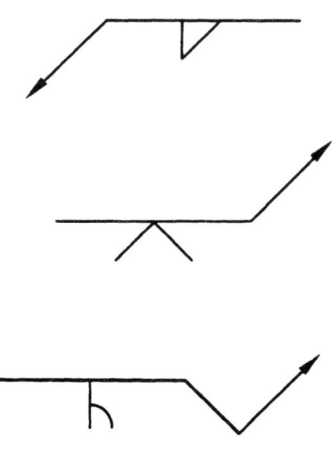

3.2.2 Other Side. Welds on the other side of the joint shall be specified by placing the weld symbol above the reference line (see 3.1.1).

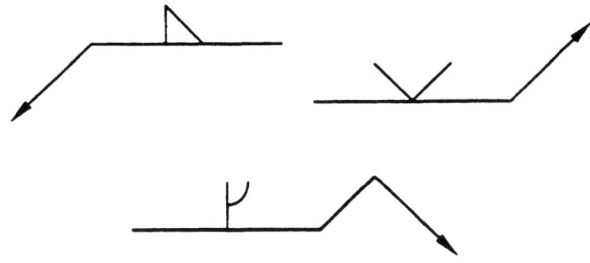

3.2.3 Both Sides. Welds on both sides of the joint shall be specified by placing weld symbols both below and above the reference line.

3.2.3.1 Symmetrical Weld Symbols. If the weld symbols used, on both sides of the reference line, have axes of symmetry that are perpendicular, or normal, to the reference line, then these axes of the symbols shall be directly aligned across the reference line. Staggered intermittent welds are an exception.

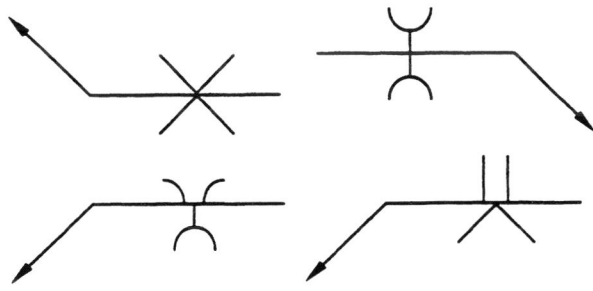

3.2.3.2 Nonsymmetrical Weld Symbols. If either of the weld symbols used lacks an axis of symmetry perpendicular, or normal, to the reference line, then the left sides of the weld symbols shall be directly aligned across the reference line. Staggered intermittent welds are an exception.

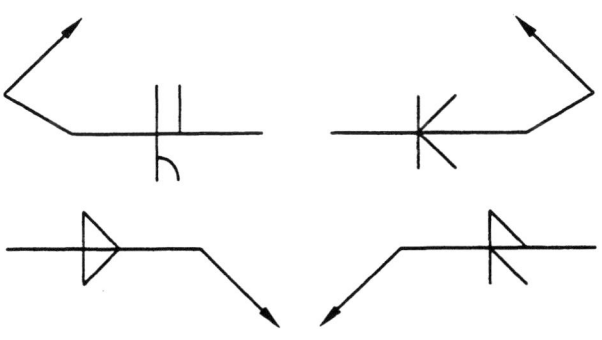

3.3 Orientation of Specific Weld Symbols. Fillet, bevel-groove, J-groove, and flare-bevel-groove weld symbols shall be drawn with the perpendicular leg always to the left.

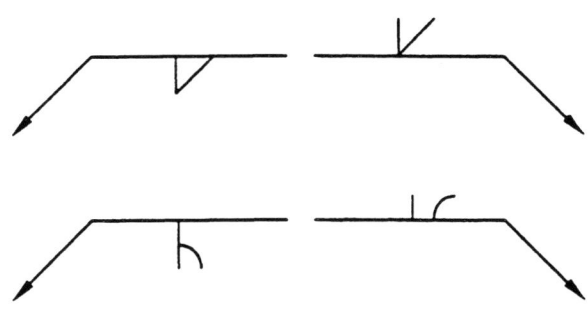

3.4 Break in Arrow. When only one joint member is to have a bevel, or J-groove, the arrow shall have a break, and point toward that member (see Figure 6). The arrow need not be broken if it is obvious which member is to have a bevel or J-groove. It shall not be broken if there is no preference as to which member is to have a bevel or J-groove.

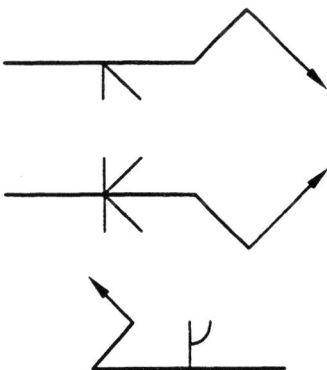

3.5 Combined Weld Symbols. For joints requiring more than one weld type, a symbol shall be used to specify each weld (see Figure 7).

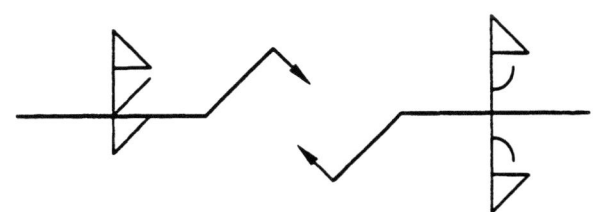

3.6 Multiple Arrow Lines. Two or more arrows may be used with a single reference line to point to locations where identical welds are specified [see Figures 9(A) and 10].

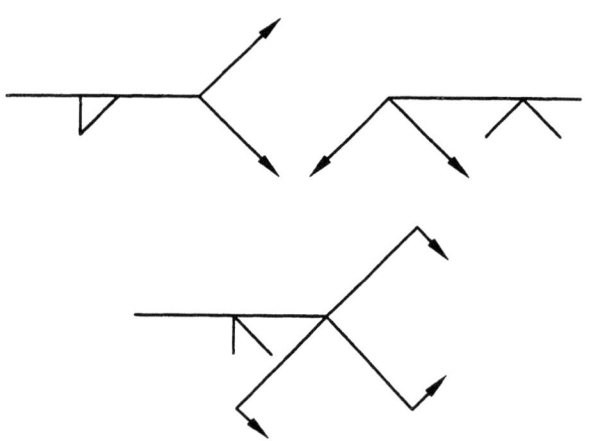

3.7 Multiple Reference Lines

3.7.1 Sequence of Operations. Two or more reference lines may be used to indicate a sequence of operations. The first operation is specified on the reference line nearest the arrow. Subsequent operations are specified sequentially on other reference lines.

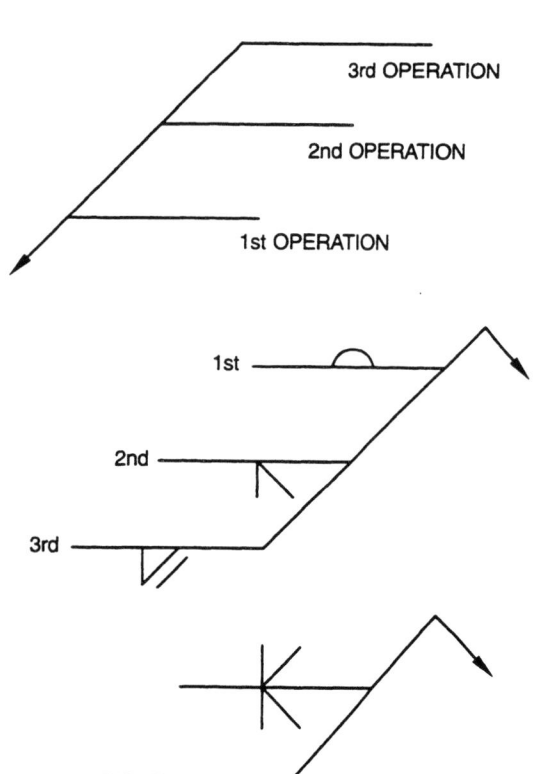

3.7.2 Supplementary Data. The tail of additional reference lines may be used to specify data supplementary to welding symbol information.

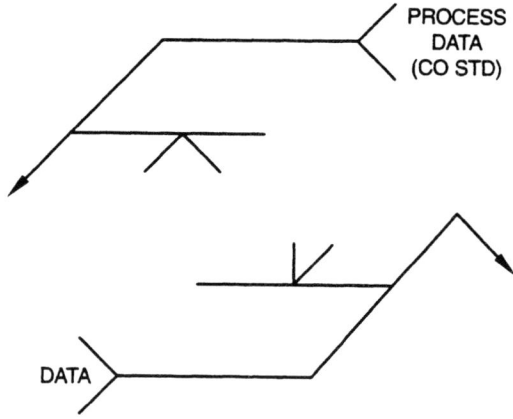

3.7.3 Field Weld and Weld All-Around Symbols. When required, the weld-(or examine-) all-around symbol shall be placed at the junction of the arrow and reference line for each operation to which it is applicable. The field weld symbol may also be applied to the same location.

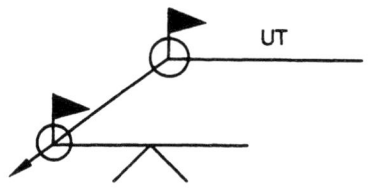

3.8 Field Weld Symbol. Field welds (welds not made in a shop or at the place of initial construction) shall be specified by adding the field weld symbol. The flag shall be placed at a right angle to, and on either side of, the reference line at the junction with the arrow (see Annex B3.8).

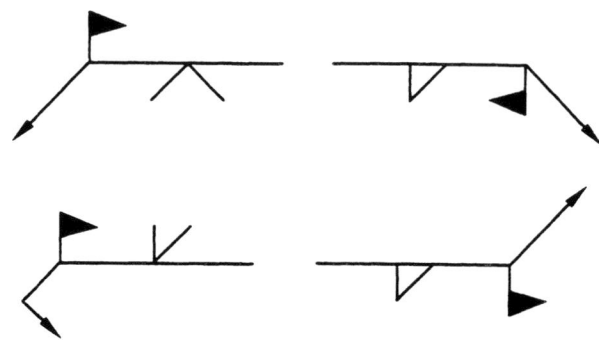

3.9 Extent of Welding Denoted by Symbols

3.9.1 Weld Continuity. Unless otherwise indicated, welding symbols shall denote continuous welds.

3.9.2 Changes in the Direction of Welding. Symbols only apply between any changes in the direction of welding, or to the extent of hatching or dimension lines (see Figure 8), except when the weld-all-around symbol is used [see Figure 9(B), (C), (D), and (E)]. Additional welding symbols or multiple arrows shall be used to specify the welds required for any changes in direction. When it is desirable to use multiple arrows on a welding symbol, the arrows shall originate from a single reference line [see Figure 9(A)] or from the first reference line in the case of a multiple reference line symbol. See Annex B3.9.2 for applications involving square and rectangular tubing.

3.9.3 Hidden Members. When the welding of a hidden member is the same as that of a visible member, it may be specified as shown below. If the welding of a hidden member is different from that of a visible member, specific information for the welding of both shall be specified. If needed for clarification, auxiliary illustrations or views shall be provided.

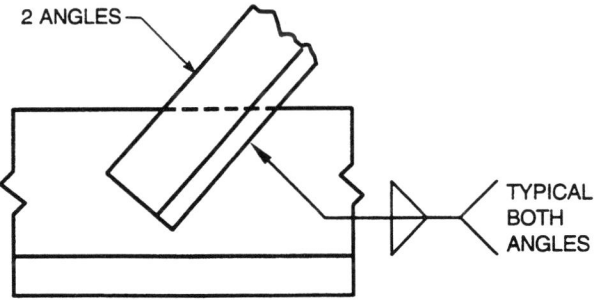

3.9.4 Weld Location Specified. A weld, with a length less than the available joint length whose location is significant, shall have the location specified on the drawing [see Figure 8(C)].

3.9.5 Weld Location Not Specified. A weld, with a length less than the available joint length and not critical regarding location, may be specified without indicating the location as shown in Figure 8(D).

3.10 Weld-All-Around Symbol

3.10.1 Welds in Multiple Directions or Planes. A continuous weld, whether single or combined type, extending around a series of connected joints may be specified by the addition of the weld-all-around symbol at the junction of the arrow and reference line. The series of joints may involve different directions and may lie in more than one plane [see Figure 9(B), (C), (D), (E), and Annex B3.10.1].

3.10.2 Circumferential Welds. Welds extending around the circumference of a pipe are excluded from the requirement regarding changes in direction and do not require the weld-all-around symbol to specify a continuous weld.

3.11 Tail of the Welding Symbol

3.11.1 Welding and Allied Process Specification. The welding and allied process to be used may be specified by placing the appropriate letter designations from Table 1 or Table 2 in the tail of the welding symbol. An auxiliary suffix from Table 4 may be used. (Tables are at the end of text.)

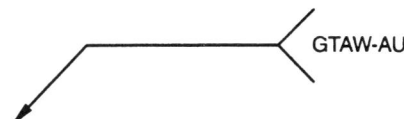

3.11.2 References. Specifications, codes or any other applicable documents may be specified by placing the reference in the tail of the welding symbol. Information contained in the referenced document need not be repeated in the welding symbol.

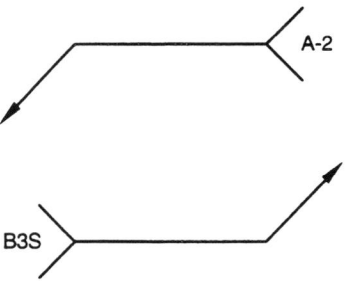

3.11.3 Welding Symbols Designated Typical. Repetitions of identical welding symbols on a drawing may be avoided by designating a single welding symbol as typical and pointing the arrow to the representative joint (see Figure 10). The user shall provide additional information to completely identify all applicable joints (see Annex B3.11.3).

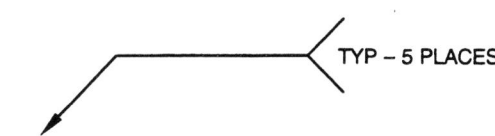

3.11.4 Designation of Special Types of Welds. When the basic weld symbols are inadequate to indicate the desired weld, the weld shall be specified by a cross section, detail, or other data with a reference thereto in the tail of the welding symbol. This may be necessary for skewed joints (see 4.13 and 5.7).

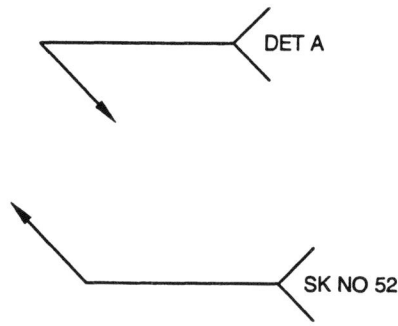

3.11.5 Omission of Tail. When no references are required, the tail may be omitted from the welding symbol.

3.11.6 Drawing Notes. Drawing notes may be used to provide information pertaining to the welds. Such information need not be repeated in the welding symbols.

3.12 Contours Obtained by Welding. Welds to be made with approximately flush, flat, convex, or concave contours without the use of mechanical finishing shall be specified by adding the flush or flat, convex, or concave contour symbol to the welding symbol.

3.13 Finishing of Welds

3.13.1 Contours Obtained by Finishing. Welds to be mechanically finished approximately flush, flat, convex, or concave shall be specified by adding the appropriate contour symbol and the finishing symbol.

3.13.2 Finishing Methods. The following finishing symbols may be used to specify the method of finishing, but not the degree of finish:

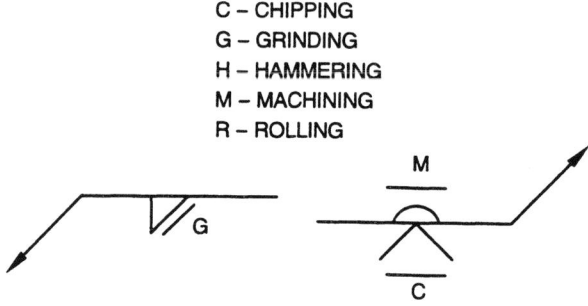

3.13.3 Finishing Method Unspecified. Welds to be finished approximately flush, flat, convex, or concave with the method unspecified shall be indicated by adding the letter "U" to the appropriate contour symbol.

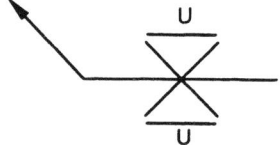

3.14 Melt-Through Symbol. The melt-through symbol shall be used only when complete joint penetration plus visible root reinforcement is required in welds made from one side (see Figure 11).

3.14.1 Melt-Through Symbol Location. The melt-through symbol shall be placed on the side of the reference line opposite the weld symbol (see Figure 11).

3.14.2 Melt-Through Dimensions. The height of root reinforcement may be specified by placing the required dimension to the left of the melt-through symbol (see Figure 11). The height of root reinforcement may be unspecified.

3.15 Melt-Through with Edge Welds

3.15.1 Melt-Through with Edge Welds on Flanged Butt Joints. Edge welds requiring complete joint penetration shall be specified by the edge weld symbol with the melt-through symbol placed on the opposite side of the reference line. The details of the flanges are considered part of the drawing and not specified by the welding symbol [See Figure 11(D)].

3.15.2 Melt-Through with Edge Welds on Flanged Corner Joints. Edge welds requiring complete joint penetration shall be specified by the edge weld symbol with the melt-through symbol placed on the opposite side of the reference line. The details of the flange are considered part of the drawing and not specified by the welding symbol [See Figure 11(E)].

3.16 Method of Drawing Symbols. Symbols may be drawn mechanically, electronically or freehand. Symbols intended to appear in publications or to be of high precision should be drawn with dimensions and proportions given in Annex A or Annex AM.

3.17 U.S. Customary and Metric Units. The same system that is the standard for the drawings shall be used on welding symbols. Dual units shall not be used on welding symbols. If it is desired to show conversions from metric to U.S. customary, or vice versa, a table of conversions may be included on the drawing. For guidance in drafting standards, reference is made to ANSI Y14, *Drafting Manual*. For guidance on the use of metric (SI) units, reference is made to ANSI/AWS A1.1, *Metric Practice Guide for the Welding Industry*.

3.18 Weld Dimension Tolerance. When a tolerance is applicable to a weld symbol dimension, it shall be shown in the tail of the welding symbol with reference to the dimension to which it applies, or the tolerance shall be specified by a drawing note, code, or specification.

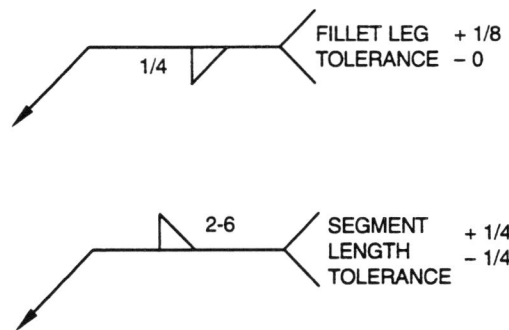

3.19 Changes in Joint Geometry During Welding. A single-reference-line welding symbol is intended to specify the joint geometry to be established prior to the start of welding. Changes in the joint geometry of groove welds resulting from the specified welding operations, such as backgouging and backing welds, are not to be included as a part of the welding symbol (see Annex B3.19).

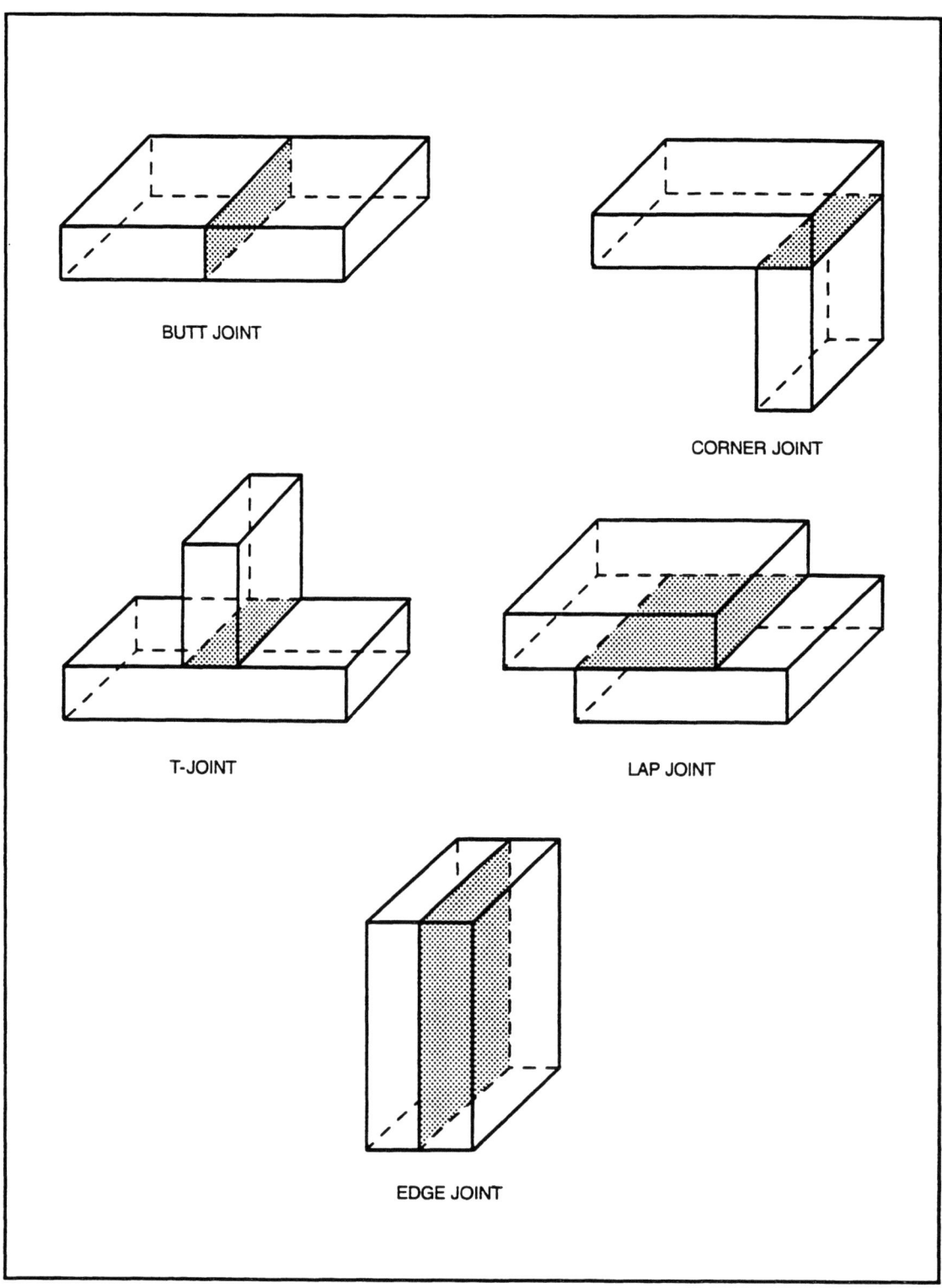

Figure 4—Basic Joints

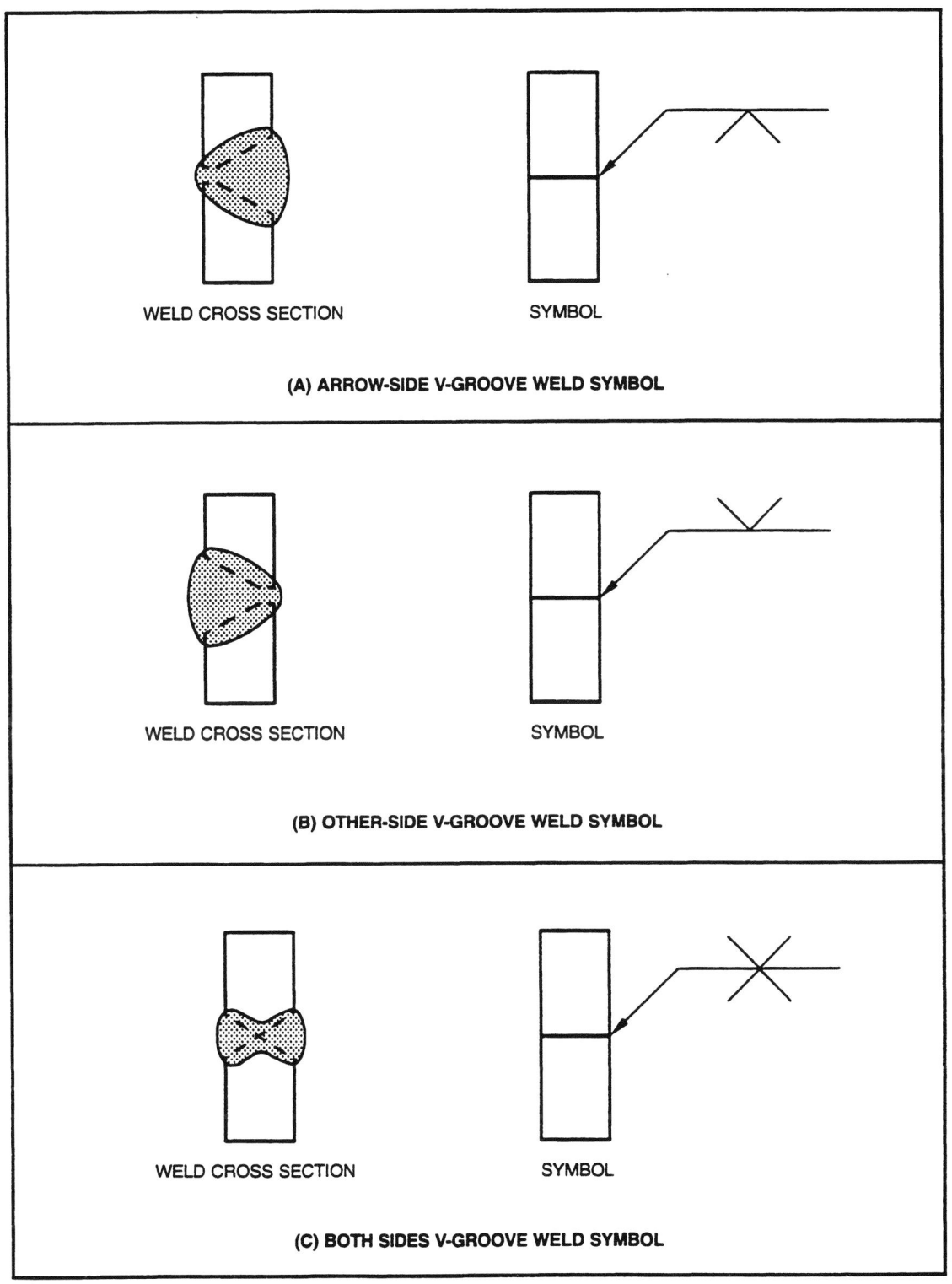

Figure 5—Applications of Arrow and Other Side Convention

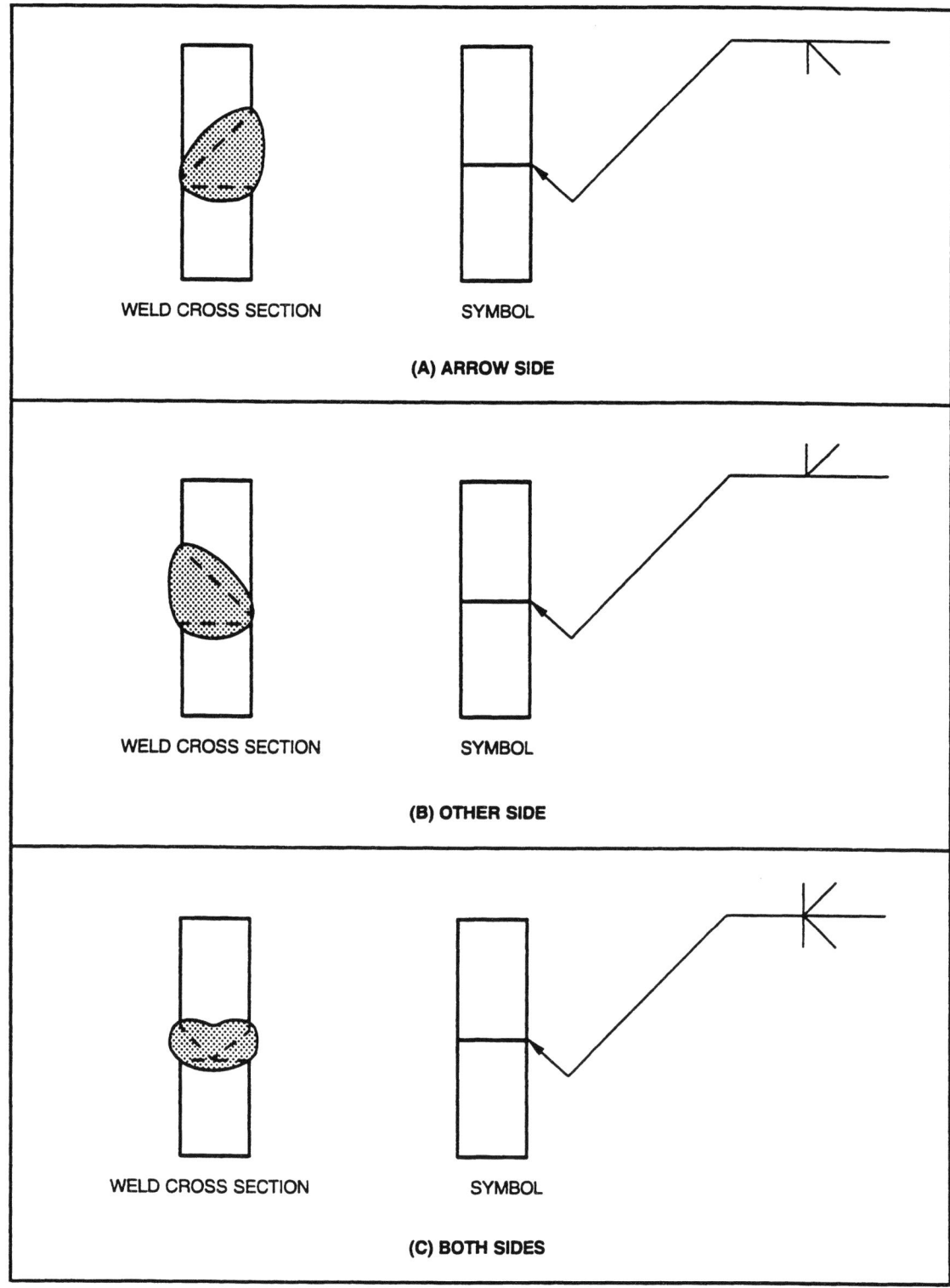

Figure 6—Applications of Break in Arrow of Welding Symbol

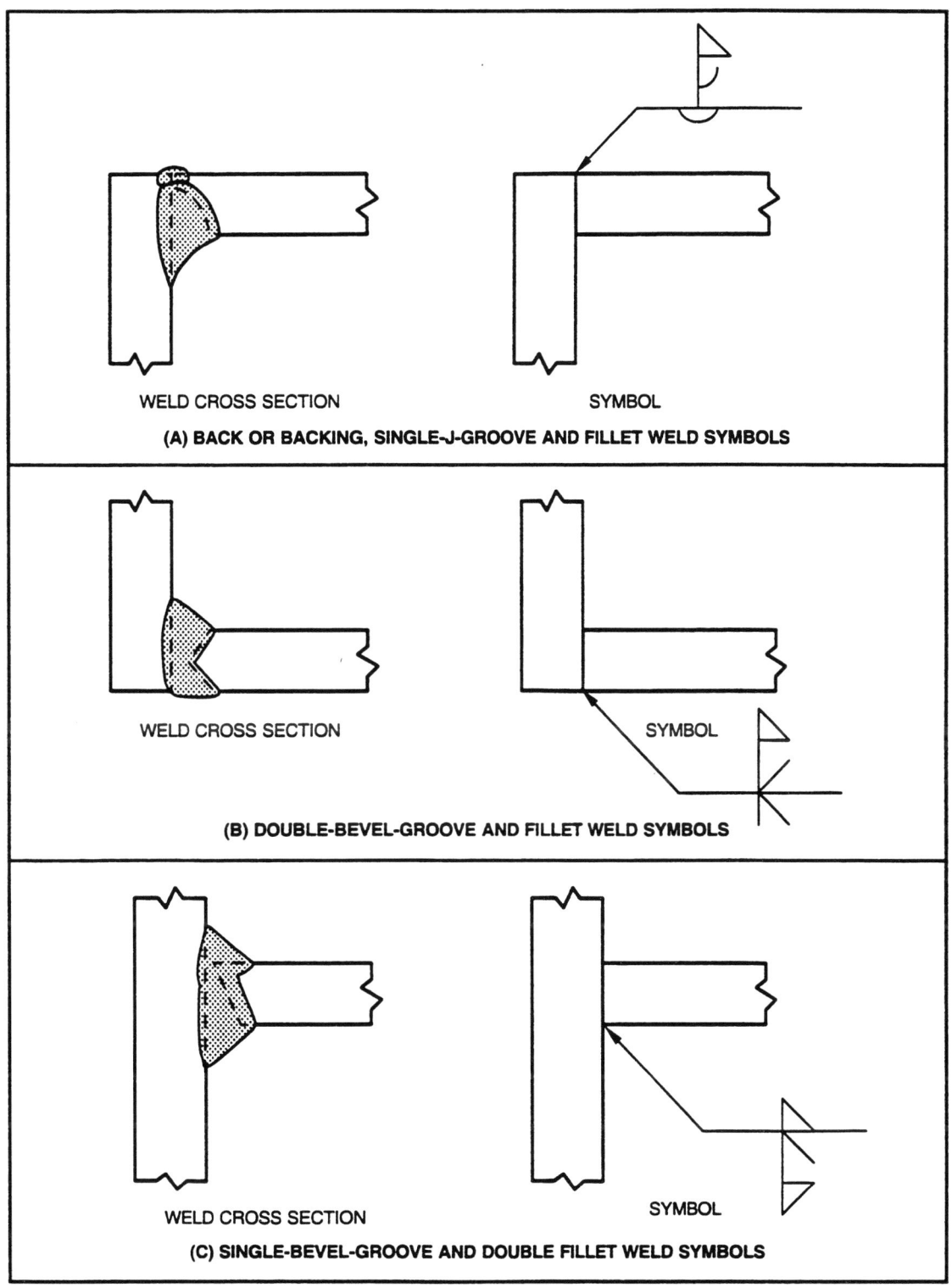

Figure 7—Combinations of Weld Symbols

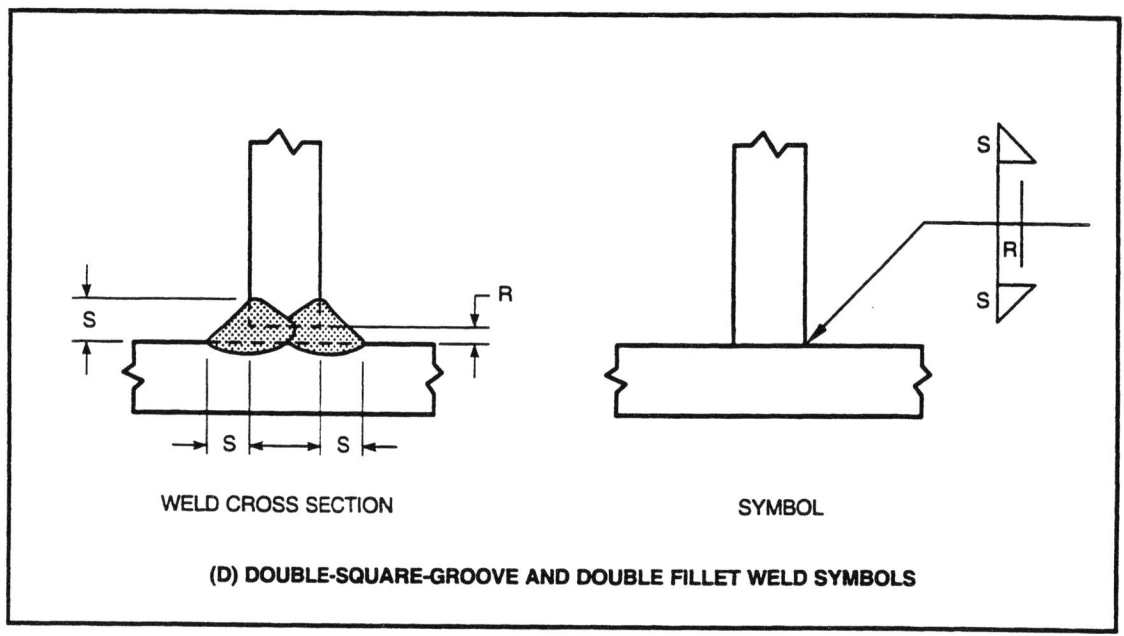

Figure 7 (Continued)—Combinations of Weld Symbols

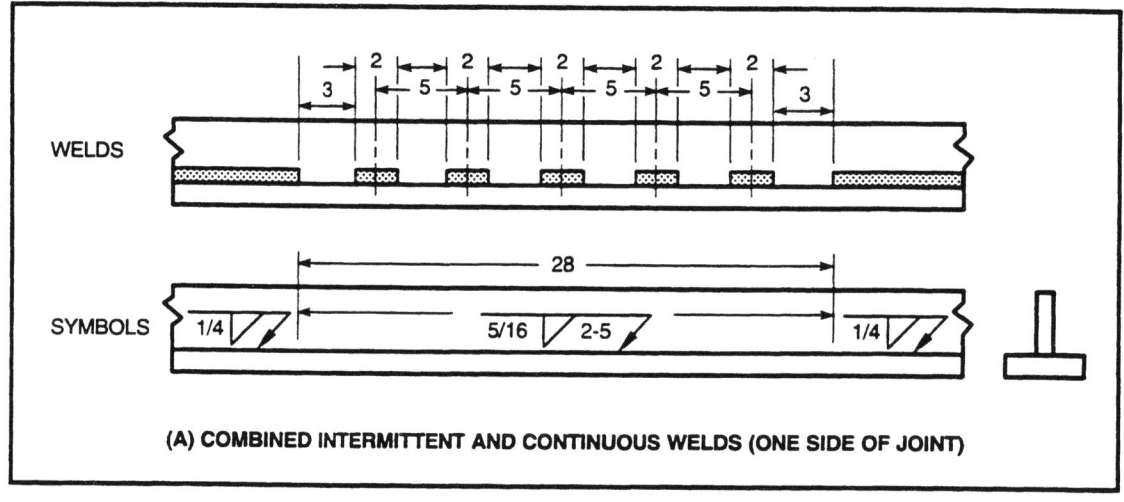

Figure 8—Specification of Location and Extent of Fillet Welds

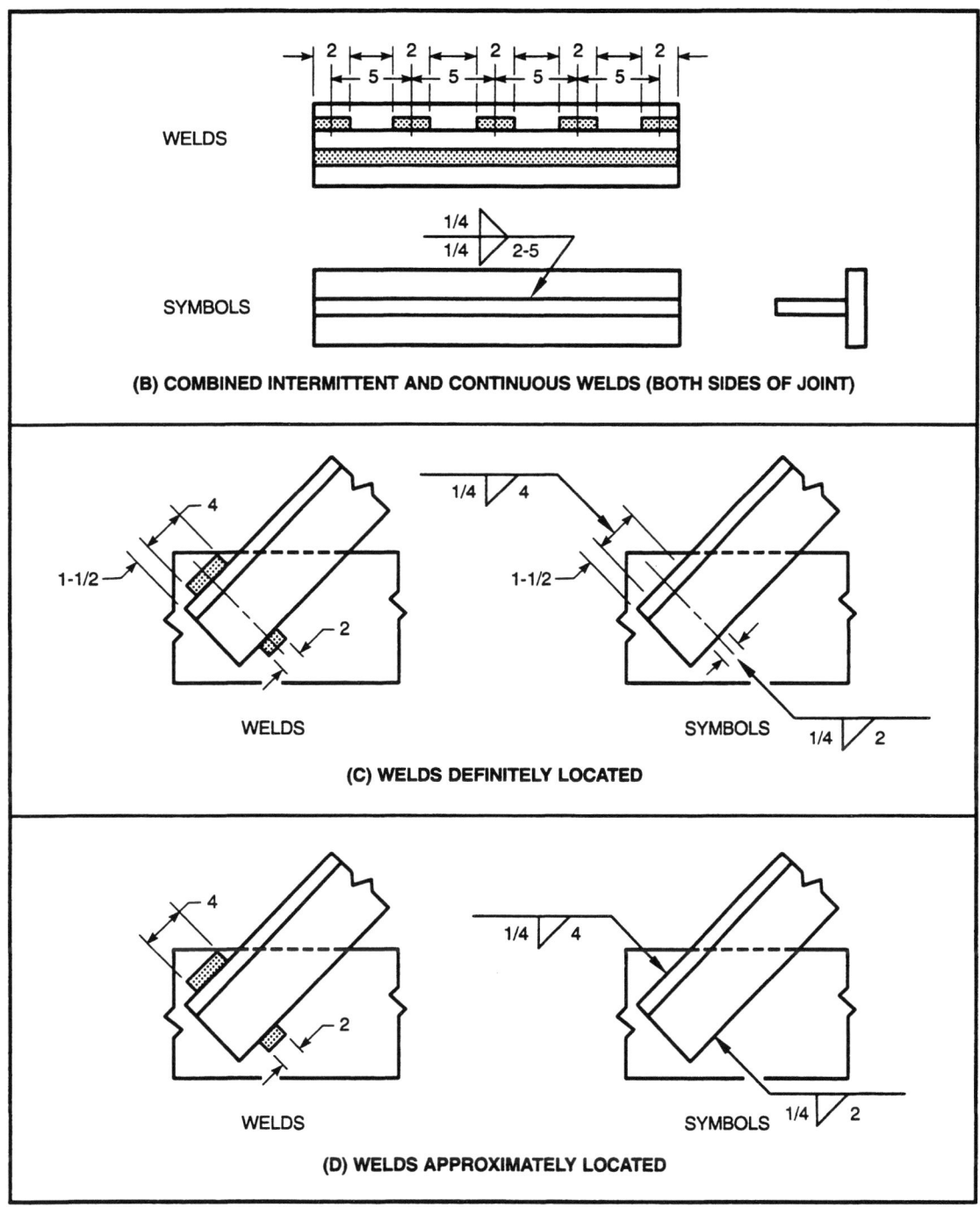

Figure 8 (Continued)—Specification of Location and Extent of Fillet Welds

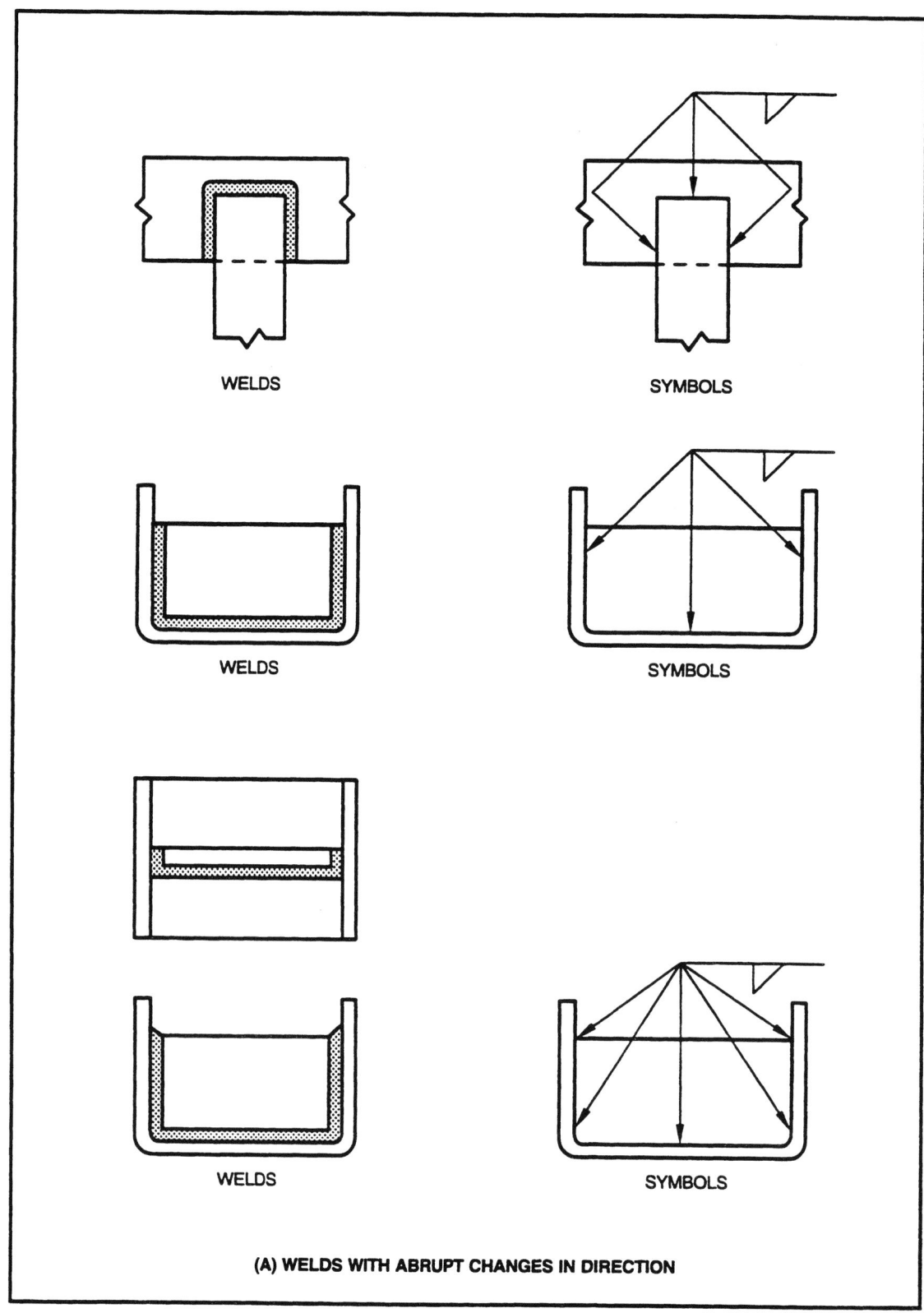

Figure 9—Specification of Extent of Welding

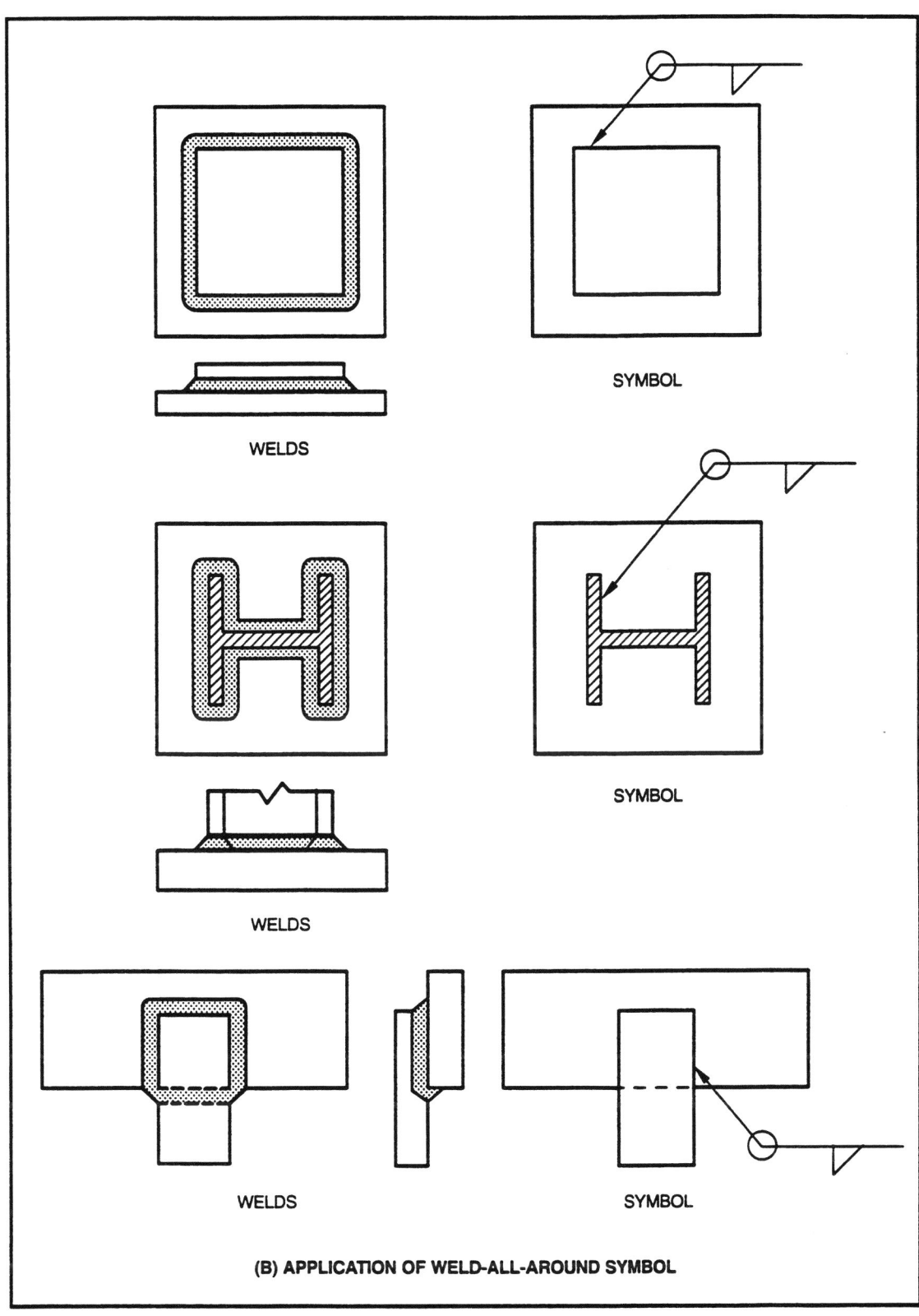

Figure 9 (Continued)—Specification of Extent of Welding

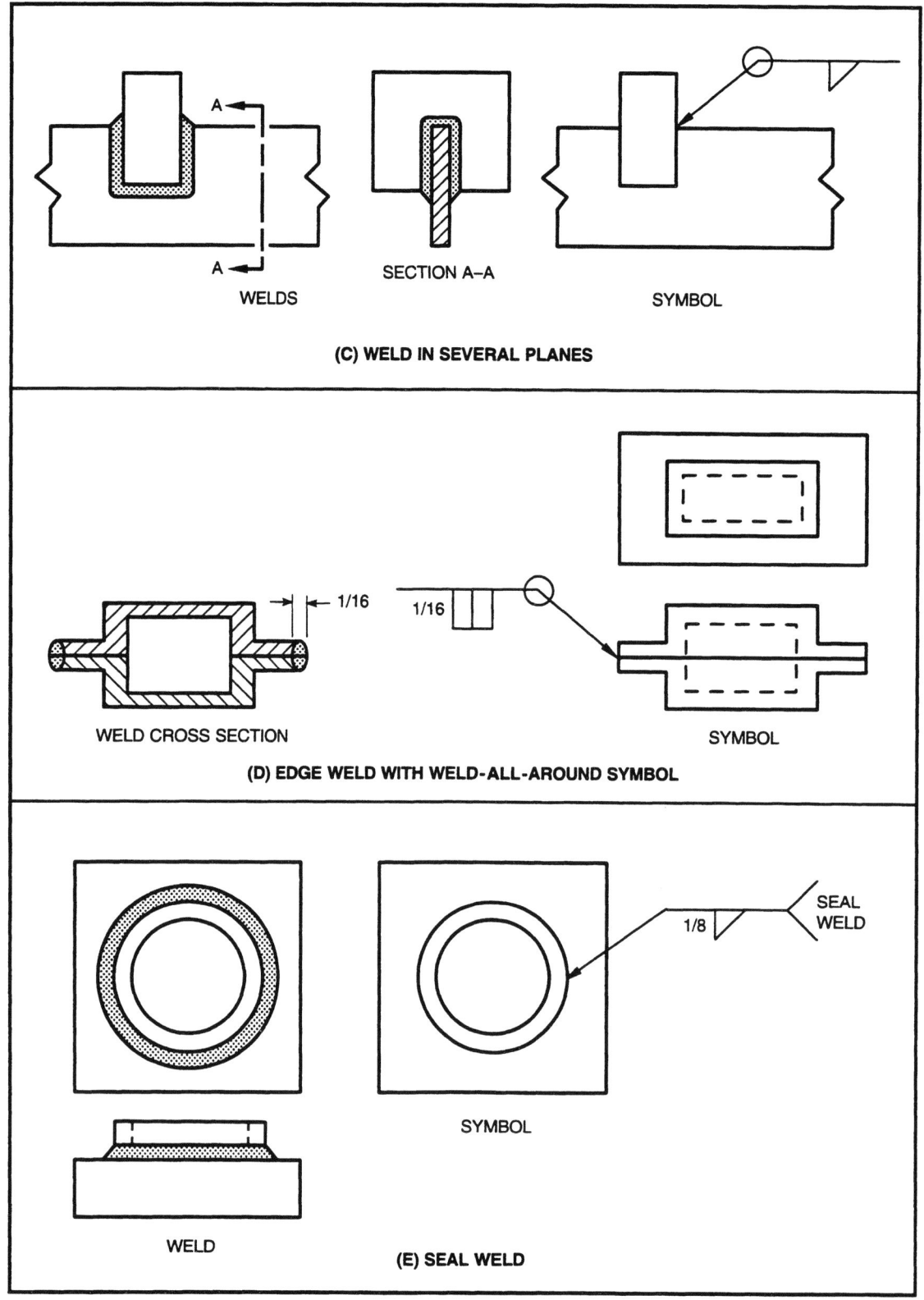

Figure 9 (Continued)—Specification of Extent of Welding

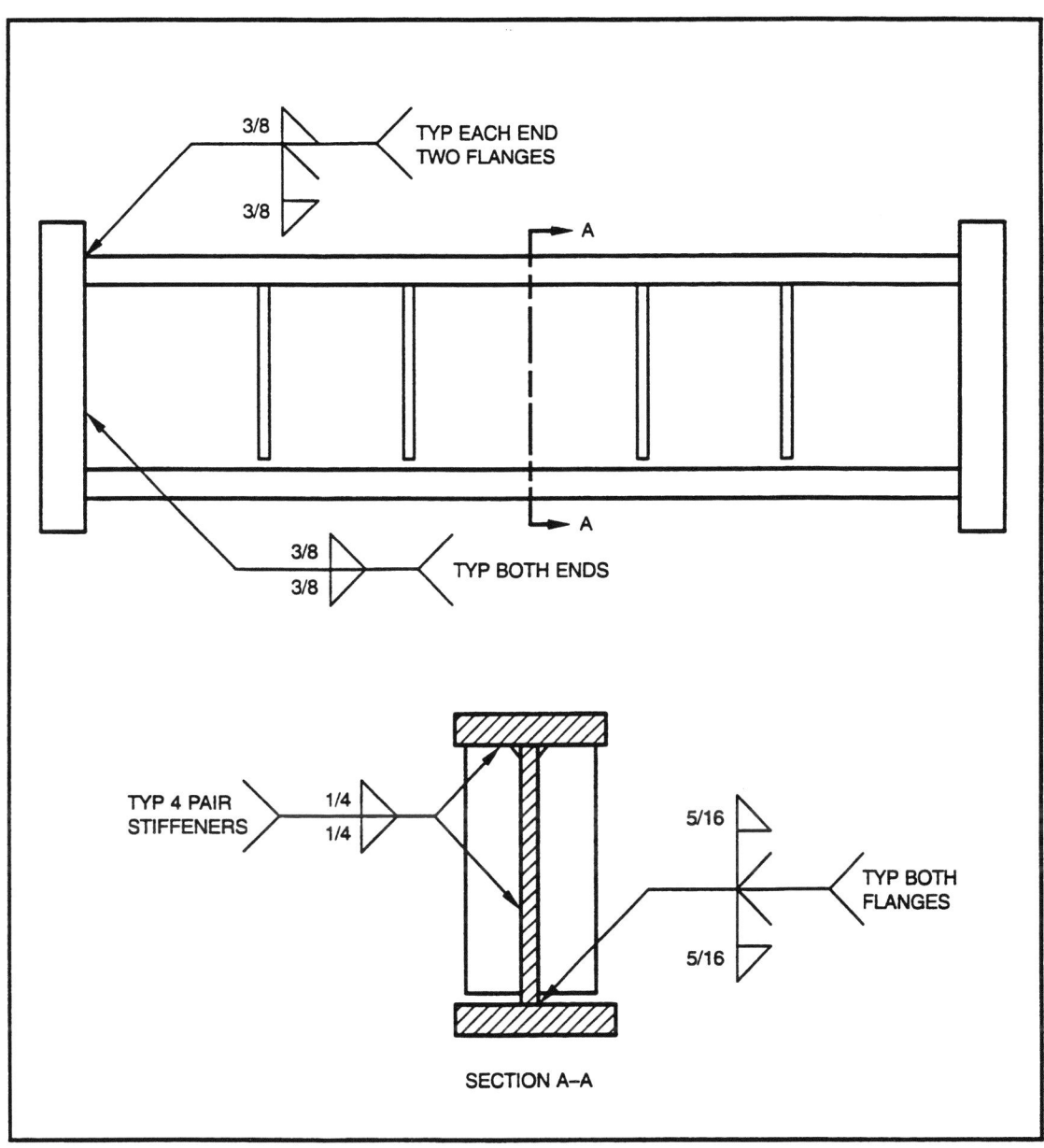

Figure 10—Applications of "Typical" Welding Symbols

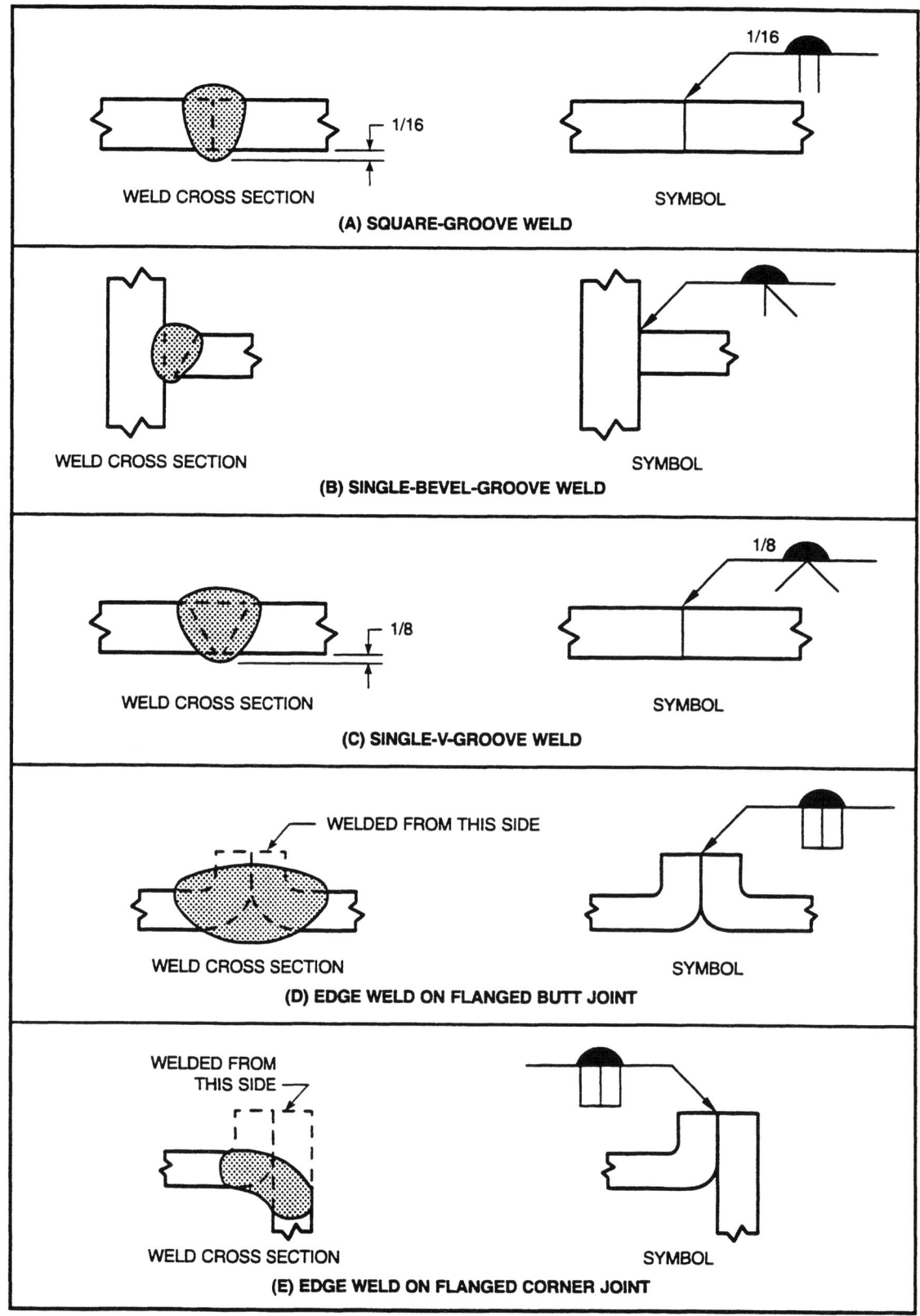

Figure 11—Applications of Melt-Through Symbol

4. Groove Welds

4.1 General

4.1.1 Single-Groove Dimensions. Groove weld dimensions shall be specified on the same side of the reference line as the weld symbol [see Figure 12(A) and (F)].

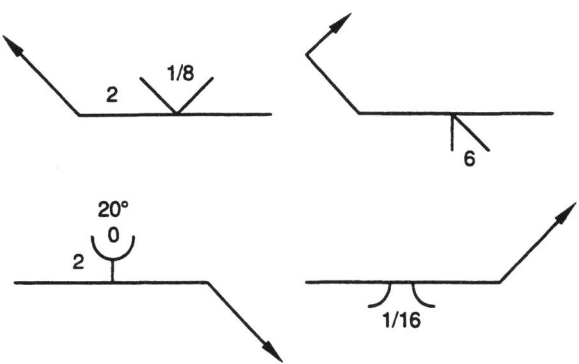

4.1.2 Double-Groove Dimensions. Each groove of a double-groove joint shall be dimensioned; however, the root opening need appear only once (see Figure 13).

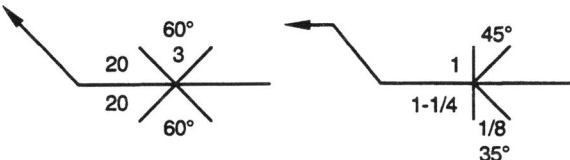

4.1.3 Broken Arrow and Straight Arrows

4.1.3.1 Broken Arrow. A broken arrow is used, when necessary, to specify which member is to have a bevel- or J-groove edge shape for single- or double-bevel and single- or double-J-groove welds (see 3.4).

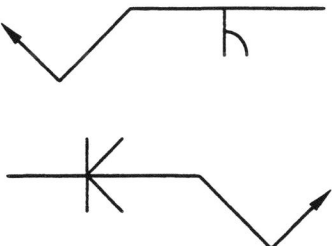

4.1.3.2 Straight Arrow for Single-Groove Welds. A straight arrow is used when either member may have the desired edge shape for single-bevel- or single-J-groove welds

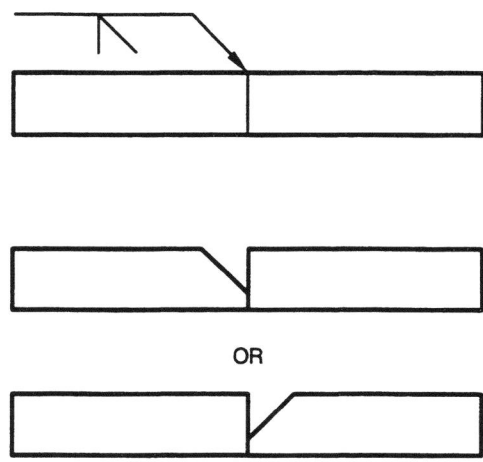

4.1.3.3 Straight Arrow for Double-Groove Welds. A straight arrow is used when either or both members may have the desired edge shape for double-bevel- or double-J-groove welds. The edge shape may be in one member on the arrow side of the joint and in the second member on the other side of the joint.

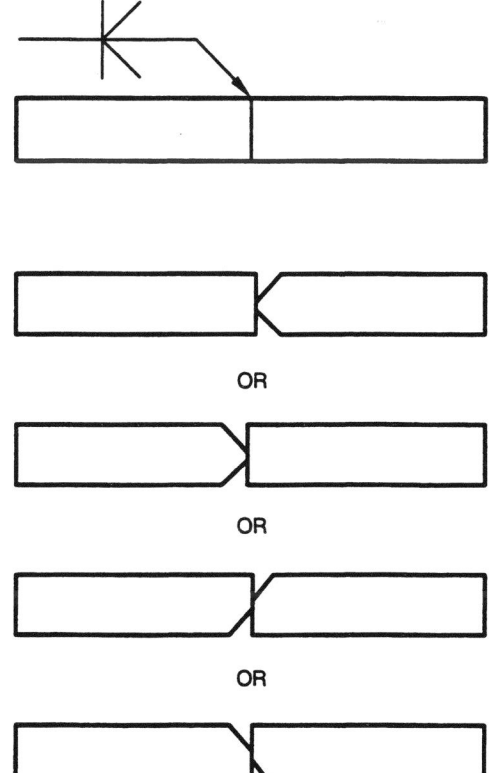

4.2 Depth of Bevel and Groove Weld Size

4.2.1 Location. The depth of bevel, S, and groove weld size, (E), shall be placed to the left of the weld symbol (see Figures 12–17).

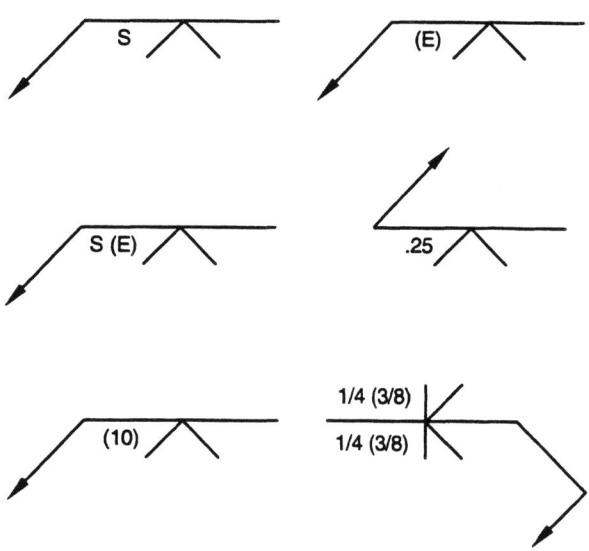

4.2.2 Complete Joint Penetration. Omitting the depth of bevel and groove weld size dimensions from the welding symbol requires complete joint penetration only for single-groove welds and double-groove welds having symmetrical joint geometry [see Figures 12(D) and (E), 21, 22(A), (B), (D), and 23 and Annex B4.2.2].

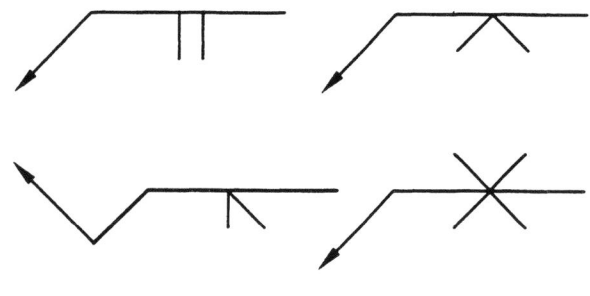

4.2.3 Partial Penetration Welds, Groove Weld Size Specified, Depth of Bevel Not Specified. The size of groove welds that extend only partly through the joint shall be specified in parentheses on the welding symbol [see Figure 12(A), (C), and (F)].

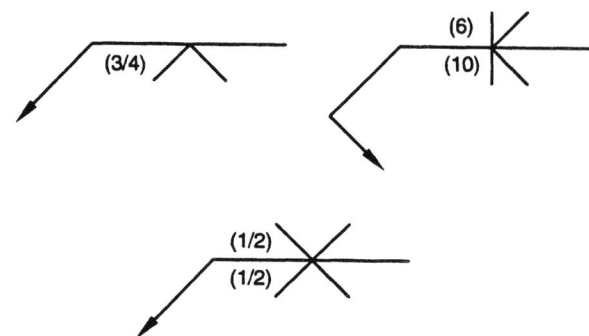

4.2.4 Complete Joint Penetration Welds, Groove Weld Size Specified, Depth of Bevel Not Specified. The size of nonsymmetrical groove welds that extend completely through the joint shall be specified in parentheses on the welding symbol (see Figure 16).

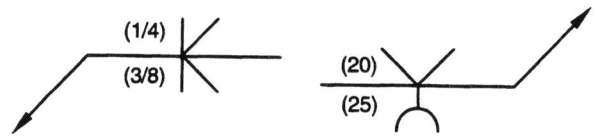

4.2.5 Depth of Bevel Specified, Groove Weld Size Specified Elsewhere. A dimension not in parentheses placed to the left of a bevel-, V-, J-, or U-groove weld symbol specifies only the depth of bevel.

4.2.6 Depth of Bevel and Groove Weld Size Specified. Except for square-groove welds, the groove weld size "(E)" in relation to the depth of bevel "S" is shown as "S(E)" to the left of the weld symbol. "(E)" only is shown for the square-groove weld (see Figures 14, 15, 17, and 20).

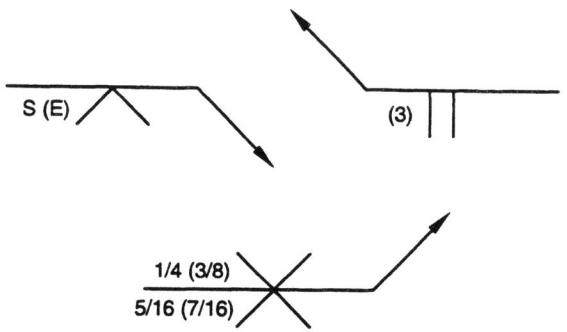

4.2.7 Depth of Bevel Specified, Groove Weld Size Not Specified. A welding symbol with a depth of bevel specified, and the groove weld size not included and not specified elsewhere, may be used to specify a groove weld size not less than the depth of bevel.

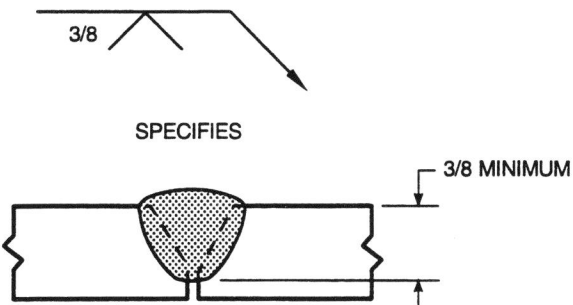

4.2.8 Joint Geometry Not Specified, Complete Joint Penetration Required. Optional joint geometry with complete joint penetration required is specified by placing the letters "CJP" in the tail of the welding symbol and omitting the weld symbol (see Figure 18).

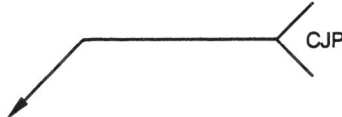

4.2.9 Joint Geometry Not Specified, Groove Weld Size Specified. For optional joint geometry, the groove weld size is specified by placing the dimension "(E)" on the arrow side or other side of the reference line as required, but omitting the weld symbol (see Figure 19).

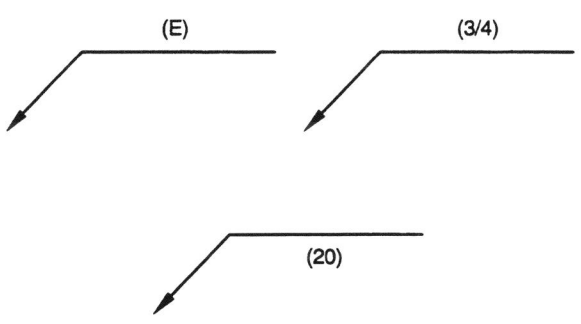

4.2.10 Flare Groove Welds. The dimension "S" of flare-groove welds is considered as extending only to the tangent point indicated below by dimension lines (see Figure 20 and Annex B4.2.9).

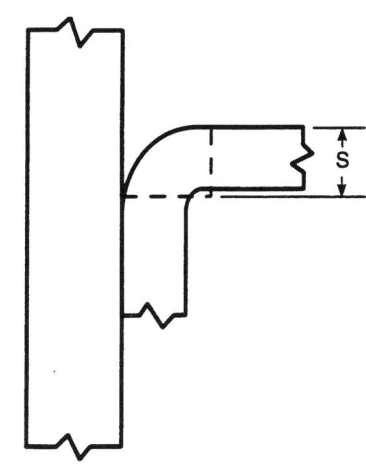
FLARE-BEVEL-GROOVE

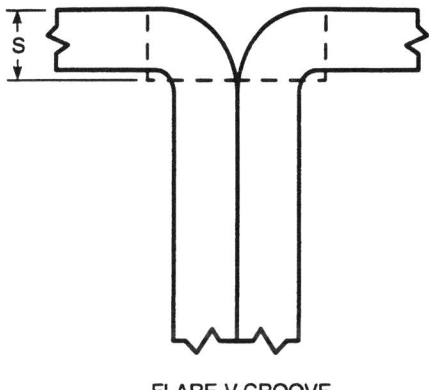
FLARE-V-GROOVE

4.3 Groove Dimensions

4.3.1 Root Opening. The root opening of groove welds shall be specified inside the weld symbol and only on one side of the reference line (see Figure 21).

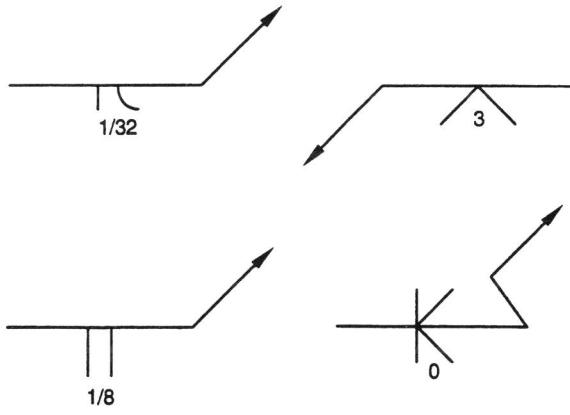

4.3.2 Groove Angle. The groove angle of groove welds shall be specified outside the weld symbol (see Figure 22).

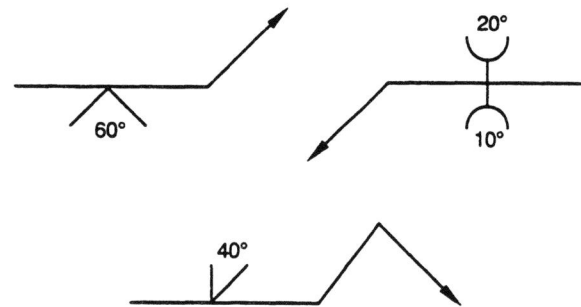

4.3.3 Radii and Root Faces.
The groove radii and root faces of U- and J-groove welds shall be specified by a cross section, detail, or other data with reference thereto in the tail of the welding symbol (see 3.11).

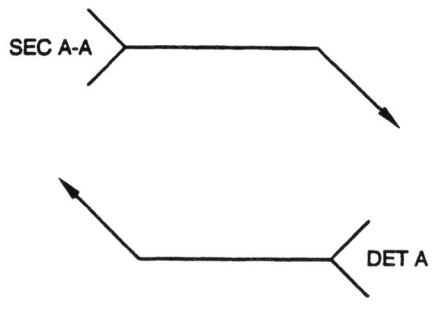

4.4 Length of Groove Welds

4.4.1 Location.
The length of a groove weld, when indicated on the welding symbol, shall be specified to the right of the weld symbol [see Figure 23(A) and (C)].

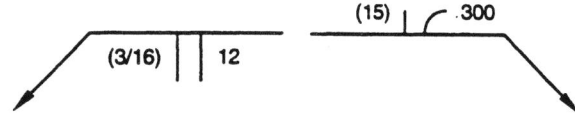

4.4.1.1 Full Length.
When a groove weld is to extend for the full length of the joint, no length dimension need be specified on the welding symbol [see Figure 23(B)].

4.4.1.2 Specific Lengths.
Specific lengths of groove welds and their locations may be specified by symbols in conjunction with dimension lines [see Figure 23(C)].

4.4.1.3 Hatching.
Hatching may be used to graphically depict groove welds.

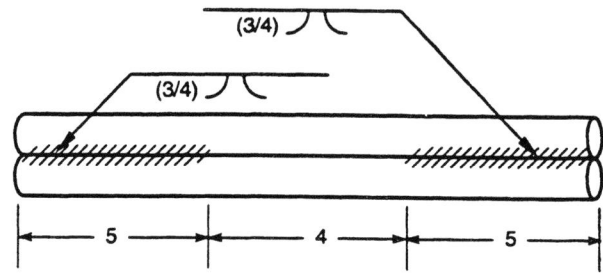

4.4.2 Changes in the Direction of Welding.
Symbols for groove welds involving changes in direction of welding shall be in accordance with 3.9.2 (see Figure 24).

4.5 Intermittent Groove Welds

4.5.1 Pitch.
The pitch of intermittent groove welds shall be the distance between the centers of adjacent weld segments on one side of the joint [see Figure 25(A)].

4.5.2 Pitch Dimension Location.
The pitch of intermittent groove welds shall be specified to the right of the length dimension following a hyphen [see Figure 25(A)].

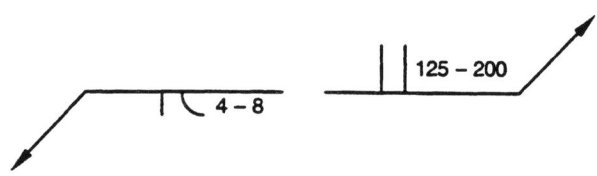

4.5.3 Chain Intermittent Groove Welds.
Dimensions of chain intermittent groove welds shall be specified on both sides of the reference line. The segments of chain intermittent groove welds shall be opposite one another across the joint [see Figure 25(B)].

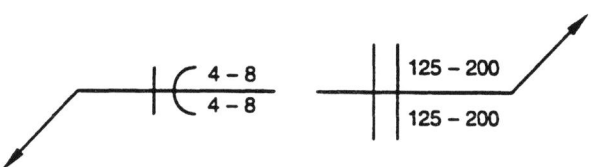

4.5.4 Staggered Intermittent Groove Welds.
Dimensions of staggered intermittent groove welds shall be specified on both sides of the reference line, and the groove weld symbols shall be offset on opposite sides of the reference line as shown below. The segments of staggered intermittent groove welds shall be symmetrically spaced on both sides of the joint as shown in Figure 25(C).

4.5.5 Extent of Welding. In the case of intermittent groove welds, additional weld lengths which are intended at the ends of the joint shall be specified by separate welding symbols and dimensioned on the drawing [see Figure 25(D)]. When no weld lengths are intended at the ends of the joint, the unwelded lengths should not exceed the clear distance between weld segments and be so dimensioned on the drawing [see Figure 25(E)].

4.5.6 Location of Intermittent Welds. When the location of intermittent welds is not obvious, such as on a circular weld joint, it will be necessary to provide specific segment locations by dimension lines (see 4.4.1.2 and 5.3.1.2) or by hatching (see 4.4.1.3 and 5.3.1.3).

4.6 Contours and Finishing of Groove Welds

4.6.1 Contours Obtained by Welding. Groove welds that are to be welded with approximately flush or convex faces without postweld finishing shall be specified by adding the flush or convex contour symbol to the welding symbol [see 3.12 and Figure 26(A)].

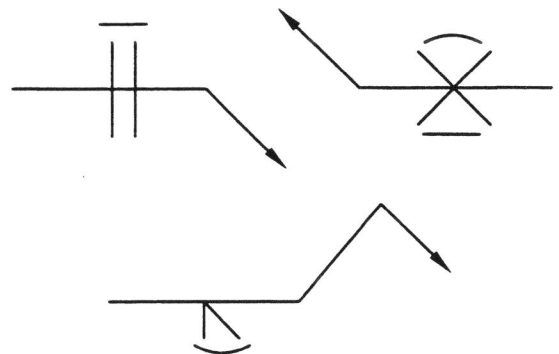

4.6.2 Contours Obtained by Postweld Finishing. Groove welds whose faces are to be finished flush or convex by postweld finishing shall be specified by adding both the appropriate contour and finishing symbols to the welding symbol. Welds that require a flat but not flush surface, require an explanatory note in the tail of the welding symbol [see 3.13 and Figure 26(B) and (C)].

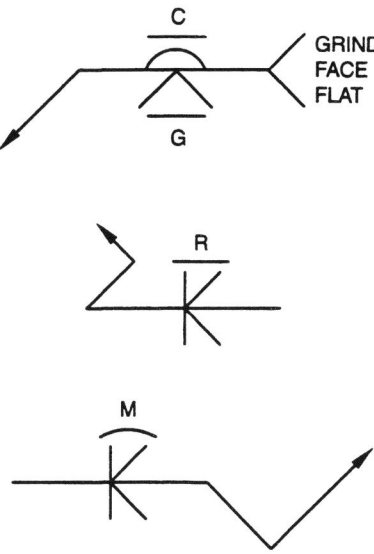

4.7 Back and Backing Welds

4.7.1 General. The back and backing weld symbols are identical. The sequence of welding determines which designation applies. The back weld is made after the groove weld, and the backing weld is made before the groove weld (see 4.7.2 and 4.7.3).

4.7.2 Back Weld Symbol. The back weld symbol is placed on the side of the reference line opposite a groove weld symbol. When a single reference line is used, "back weld" shall be specified in the tail of the welding symbol. Alternately, if multiple reference lines are used, the back weld symbol shall be placed on a reference line subsequent to the reference line specifying the groove weld [see Figure 27(A)].

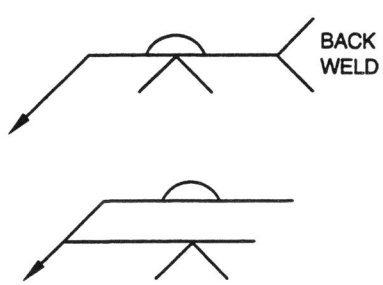

4.7.3 Backing Weld Symbol. The backing weld symbol is placed on the side of the reference line opposite a groove weld symbol. When a single reference line is used, "backing weld" shall be specified in the tail of the welding symbol. Alternately, if multiple reference lines are used, the backing weld symbol shall be placed on a reference line prior to the reference line specifying the groove weld [see Figure 27(B) and (C)].

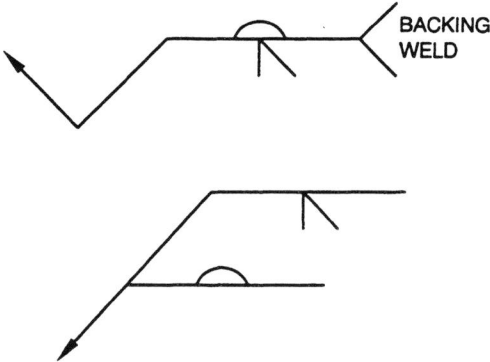

4.8 Joint with Backing. A joint with backing is specified by placing the backing symbol on the side of the reference line opposite the groove weld symbol. If the backing is to be removed after welding, an "R" shall be placed in the backing symbol [see Figure 28(A)]. Material and dimensions of backing shall be specified in the tail of the welding symbol or on the drawing.

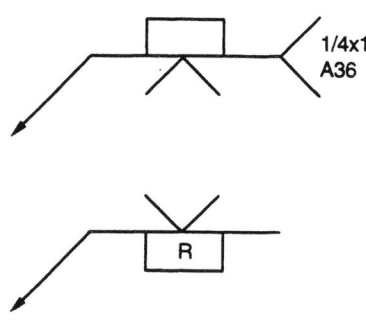

4.7.4 Contour and Finishing of Back or Backing Welds

4.7.4.1 Contours Obtained by Welding. Back or backing welds that are to be welded with approximately flush or convex faces without postweld finishing shall be specified by adding the flush or convex contour symbol to the welding symbol (see 3.12).

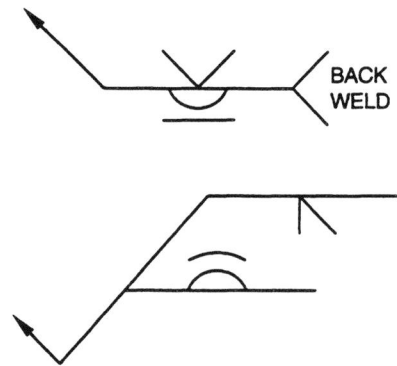

4.9 Joint with Spacer. A joint with a required spacer is specified with the groove weld symbol modified to show a rectangle within it [see Figure 28(B)]. In case of multiple reference lines, the rectangle need appear on the reference line nearest to the arrow [see Figure 28(C)]. Material and dimensions of the spacer shall be specified in the tail of the welding symbol or on the drawing.

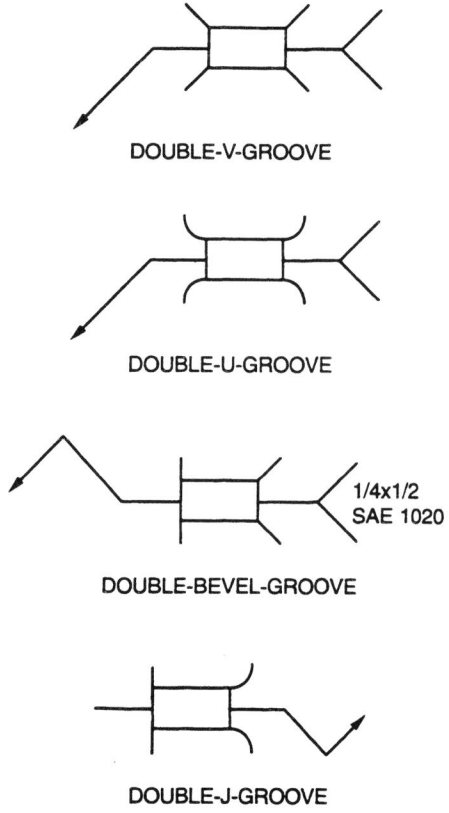

4.7.4.2 Contours Obtained by Postweld Finishing. Back or backing welds that are to be finished approximately flush or convex by postweld finishing shall be specified by adding the appropriate contour and finishing symbols to the welding symbol (see 3.13). Welds that require a flat but not flush surface, require an explanatory note in the tail of the welding symbol.

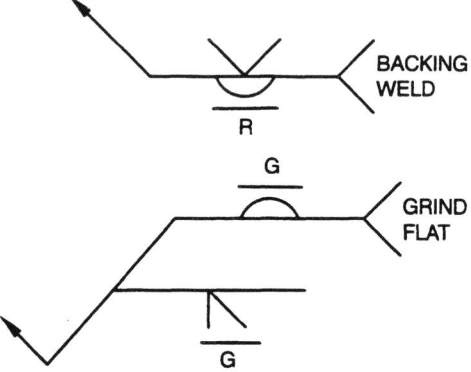

4.10 Consumable Inserts. Consumable inserts shall be specified by placing the consumable insert symbol on the side of the reference line opposite the groove weld symbol (see Figure 29). The AWS consumable insert class shall be placed in the tail of the welding symbol (for insert class see latest edition of ANSI/AWS A5.30, *Specification for Consumable Inserts*.)

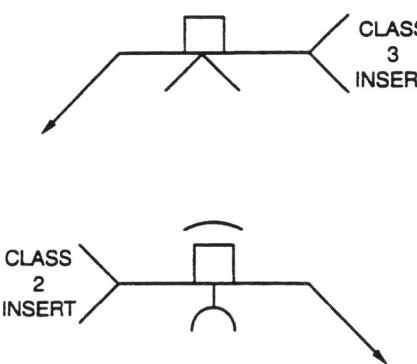

4.11 Groove Welds with Backgouging. A joint requiring complete joint penetration involving backgouging may be specified using either a single or multiple reference line welding symbol (see Figure 30). The welding symbol shall include a reference to backgouging in the tail and (1) in the case of asymmetrical double-groove welds must show the depth of bevel from each side [see Figure 28(A)], together with groove angles and root opening, or (2) in the case of single-groove welds or symmetrical double-groove welds, need not include any other information except the weld symbols [see 4.2.2 and Figure 30(B) and (C)], with groove angles and root opening.

4.12 Seal Welds. When the intent of the weld is to fulfill a sealing function only, the weld shall be specified in the tail of the welding symbol as a seal weld (see Annex B4.10).

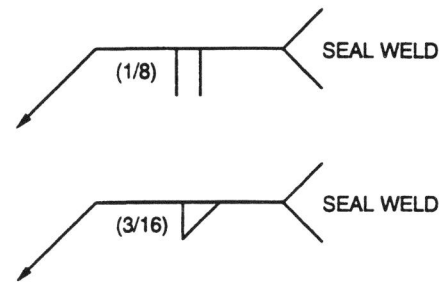

4.13 Skewed Joints. When the angle between the fusion faces is such that the identification of the weld type and, hence, proper weld symbol is in question, the detail of the desired joint and weld configuration shall be shown on the drawing with all necessary dimensions (see Figure 31).

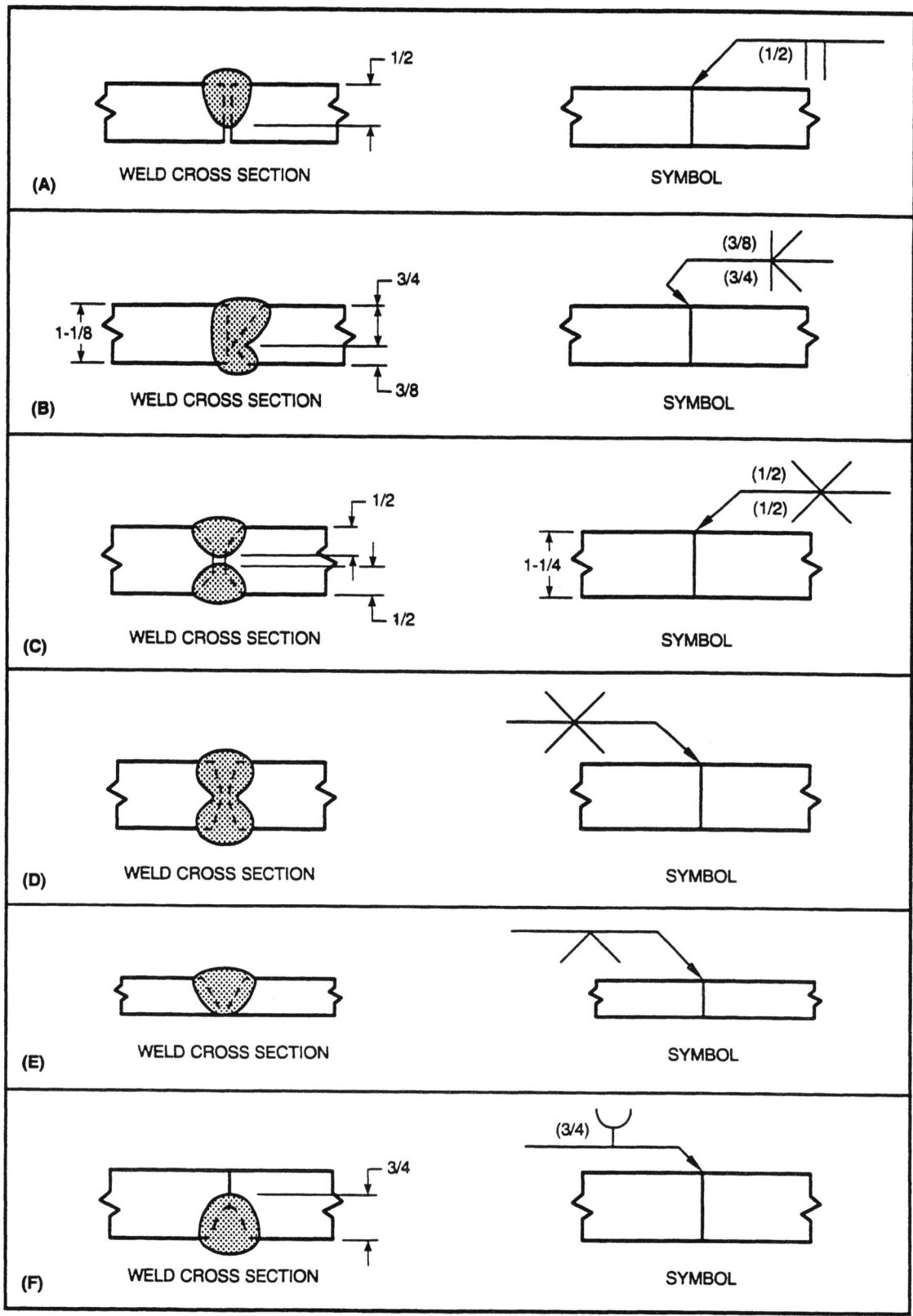

Figure 12—Specification of Groove Weld Size Depth of Bevel Not Specified

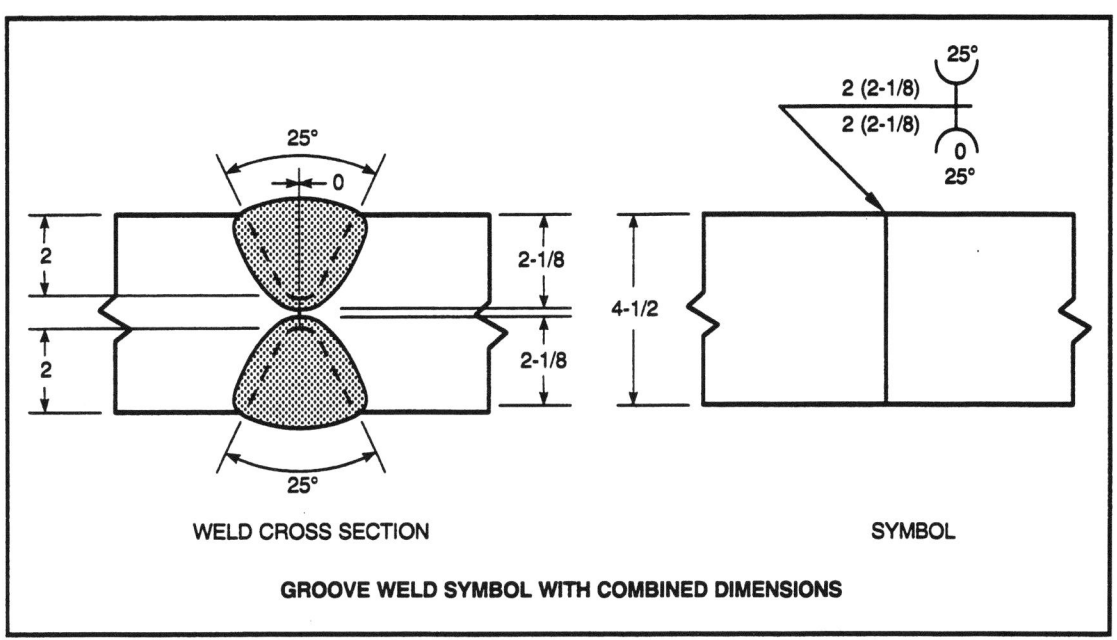

Figure 13—Application of Dimensions to Groove Weld Symbol

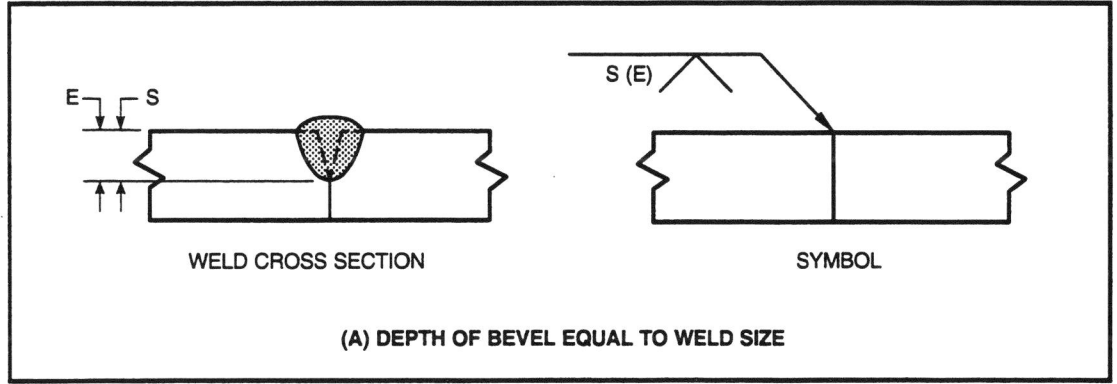

Figure 14—Groove Weld Size "(E)" Related to Depth of Bevel "S"

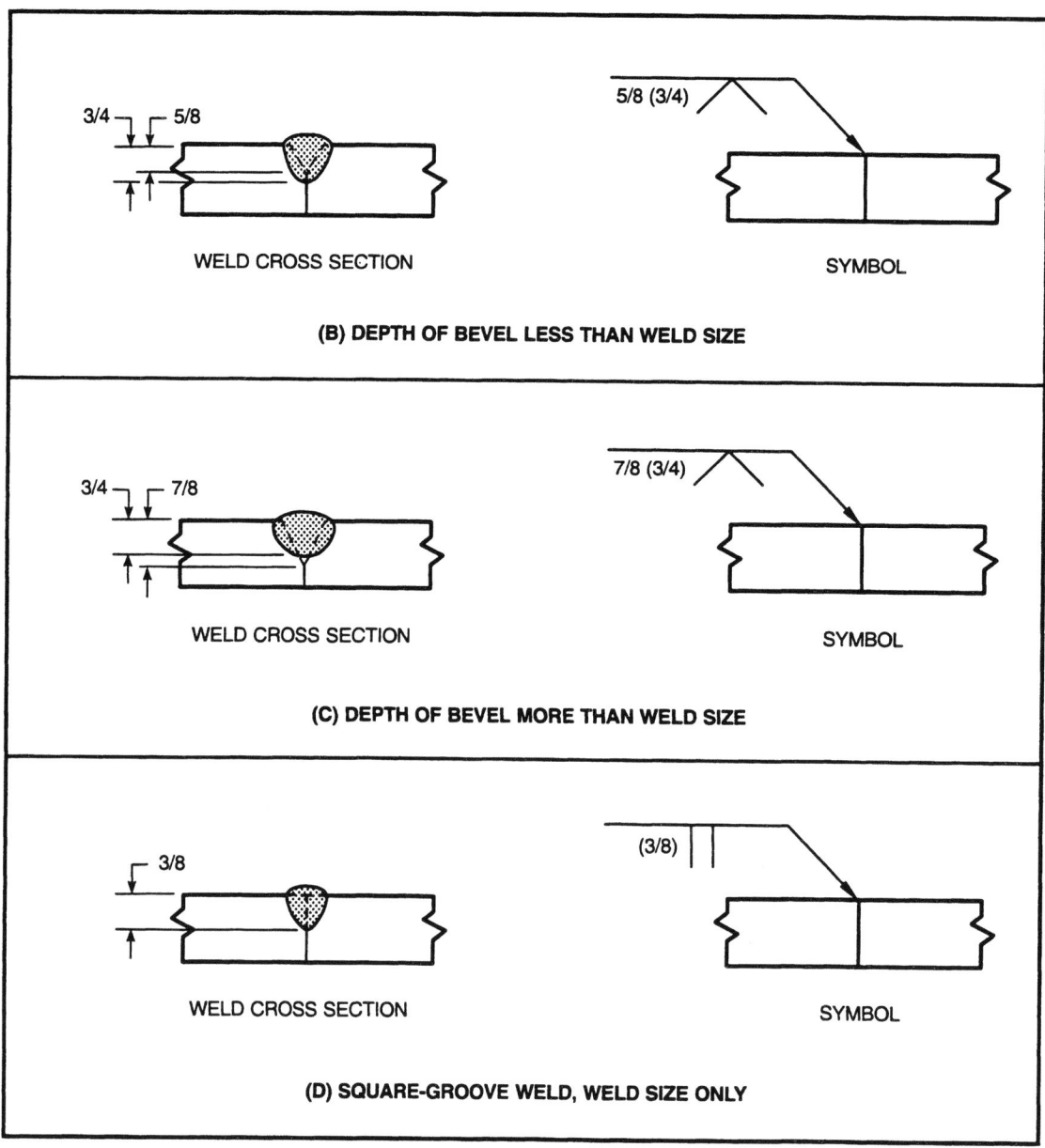

Figure 14 (Continued)—Groove Weld Size "(E)" Related to Depth of Bevel "S"

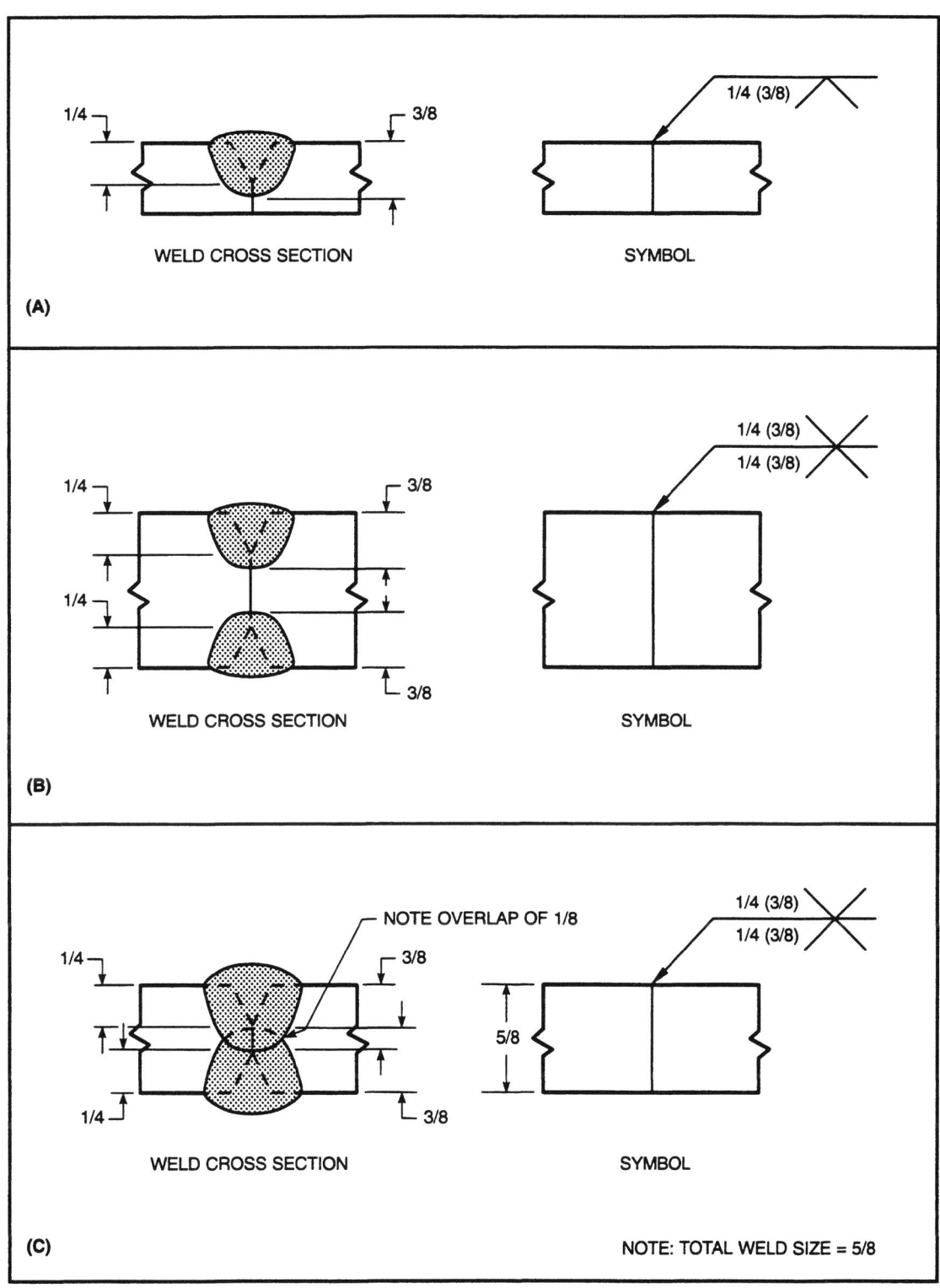

Figure 15—Specification of Groove Weld Size and Depth of Bevel

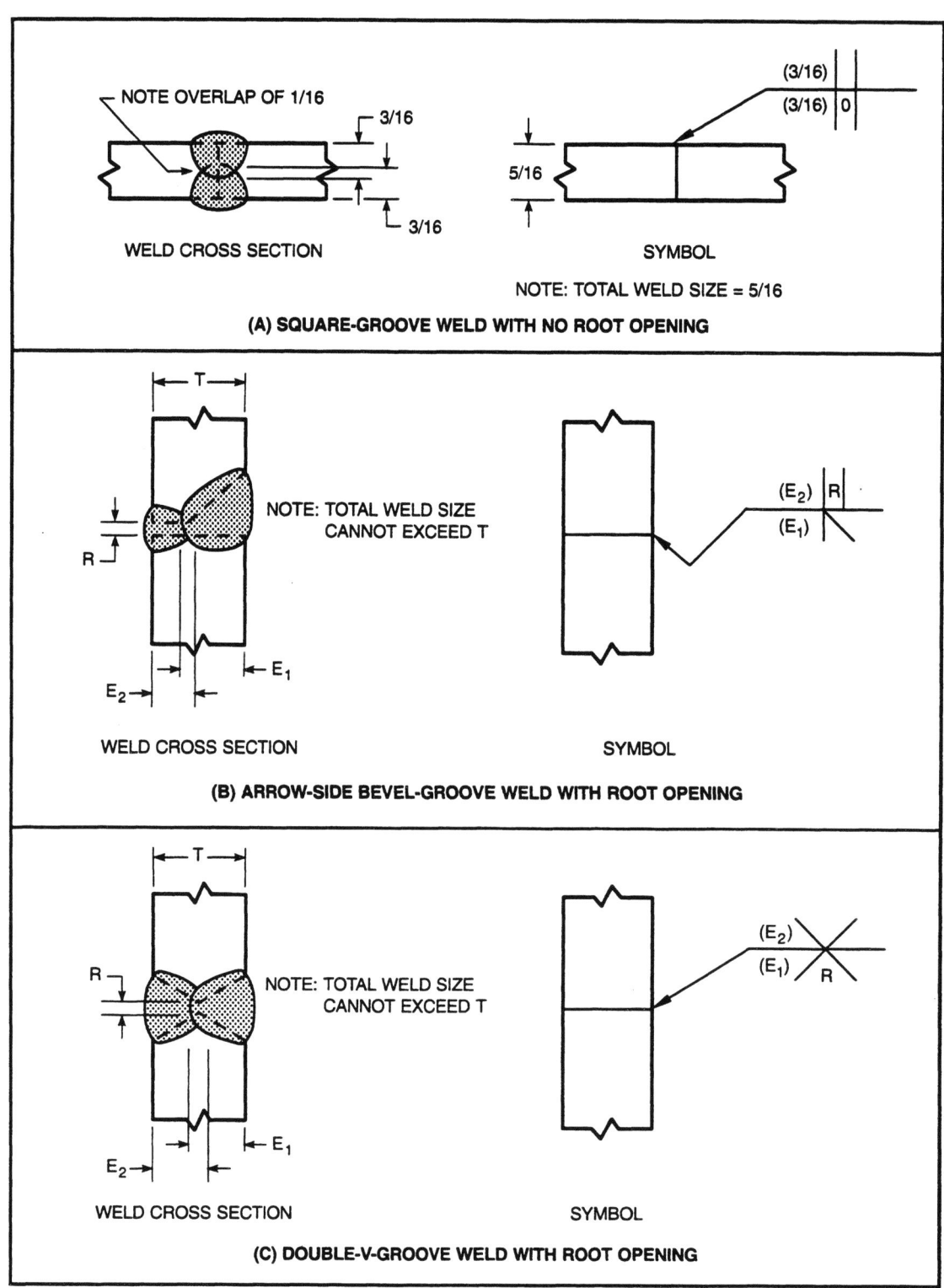

Figure 16—Specification of Groove Weld Size Only

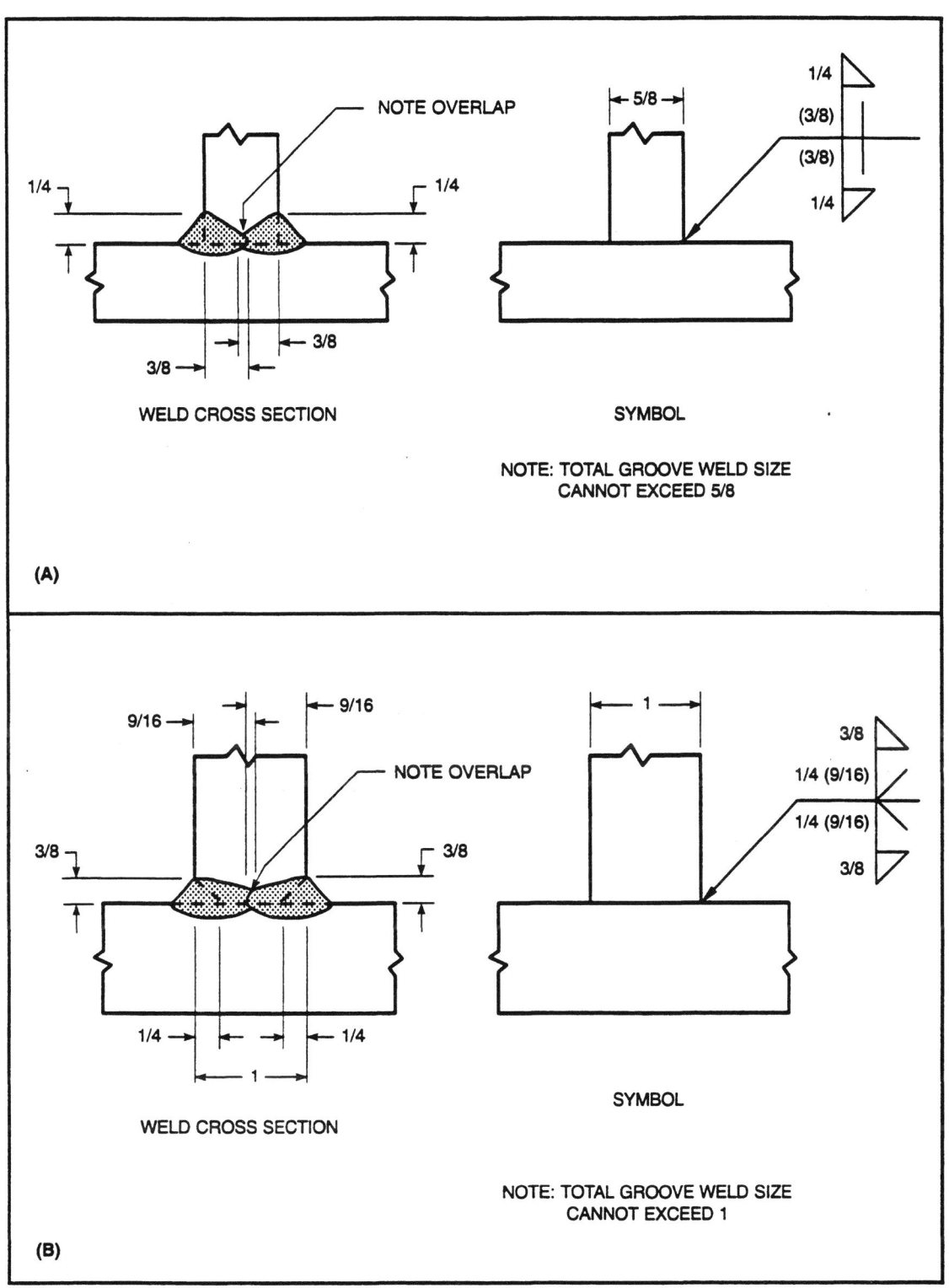

Figure 17—Combined Groove and Fillet Welds

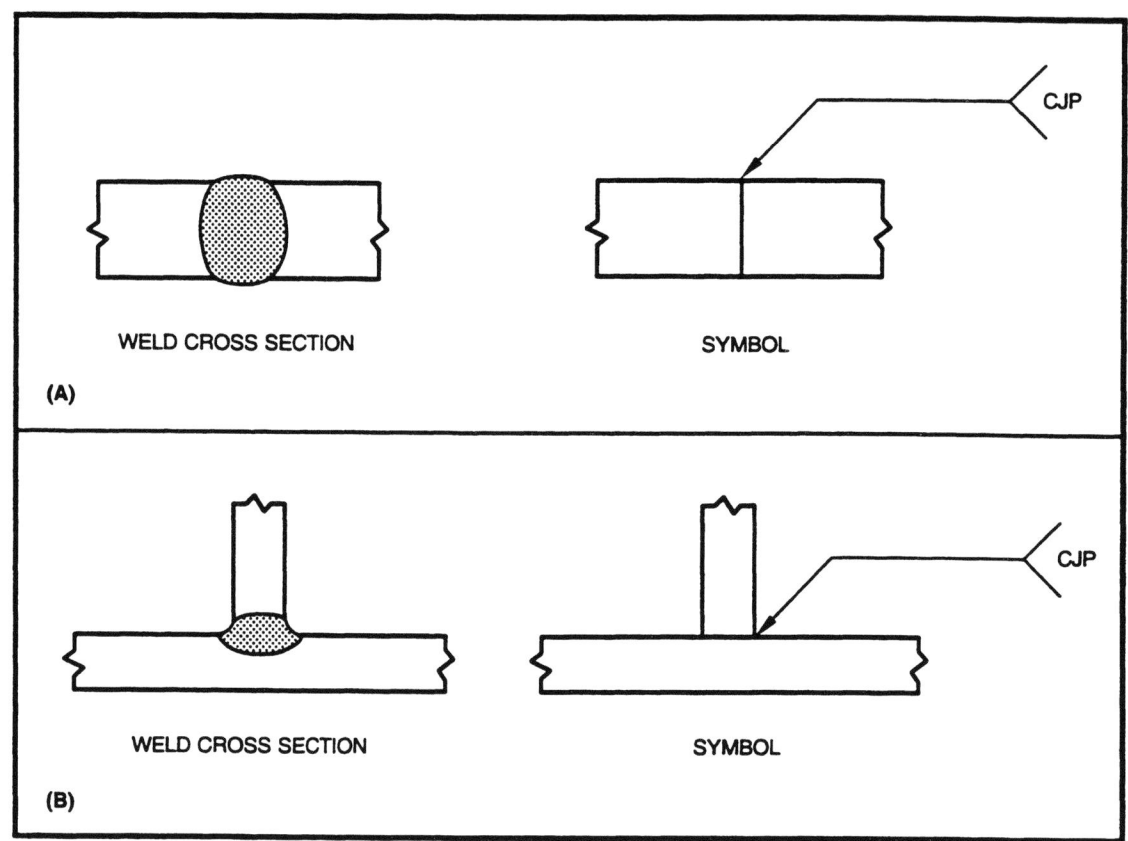

Figure 18—Complete Joint Penetration with Joint Geometry Optional

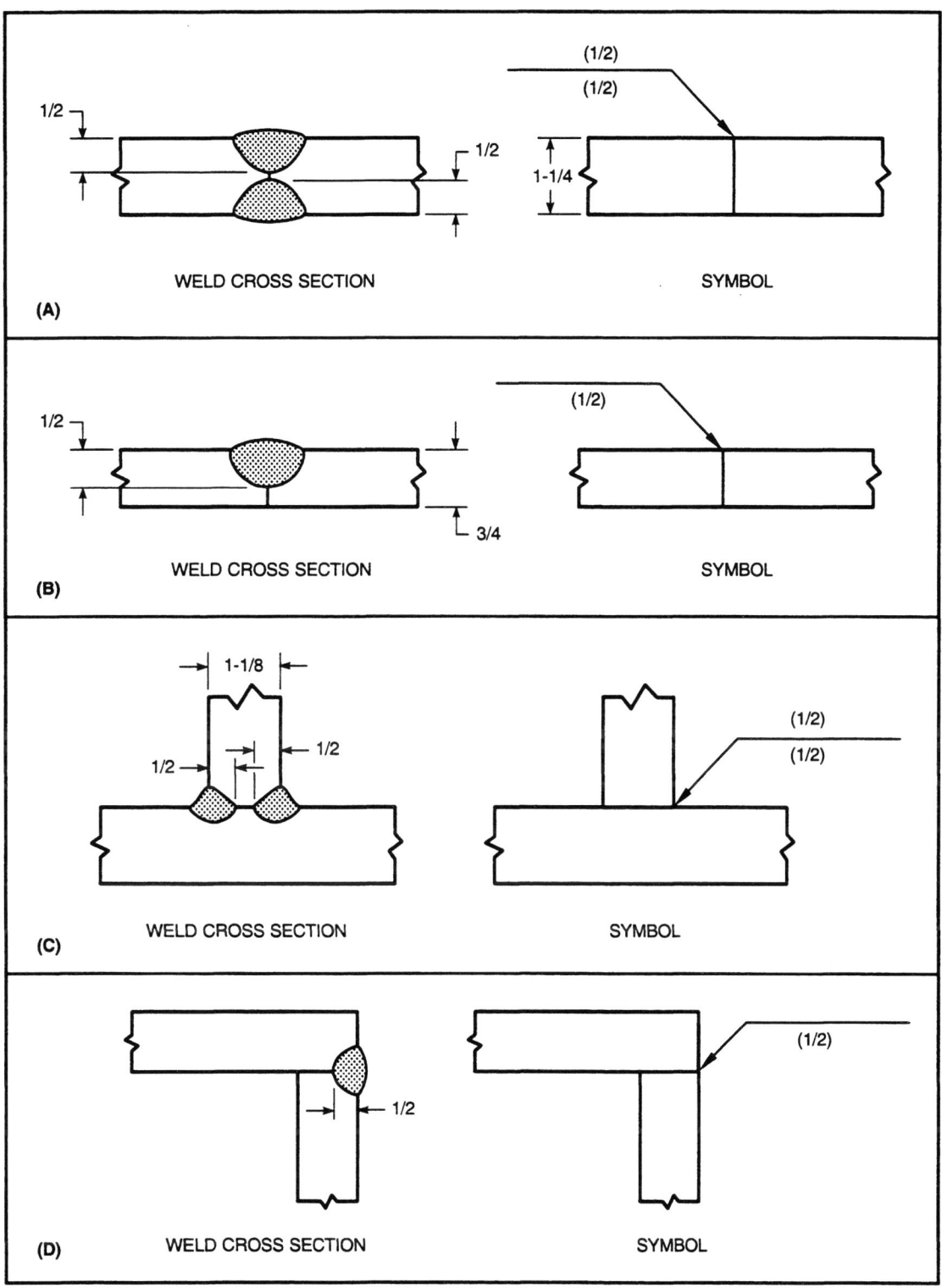

Figure 19—Partial Joint Penetration with Joint Geometry Optional

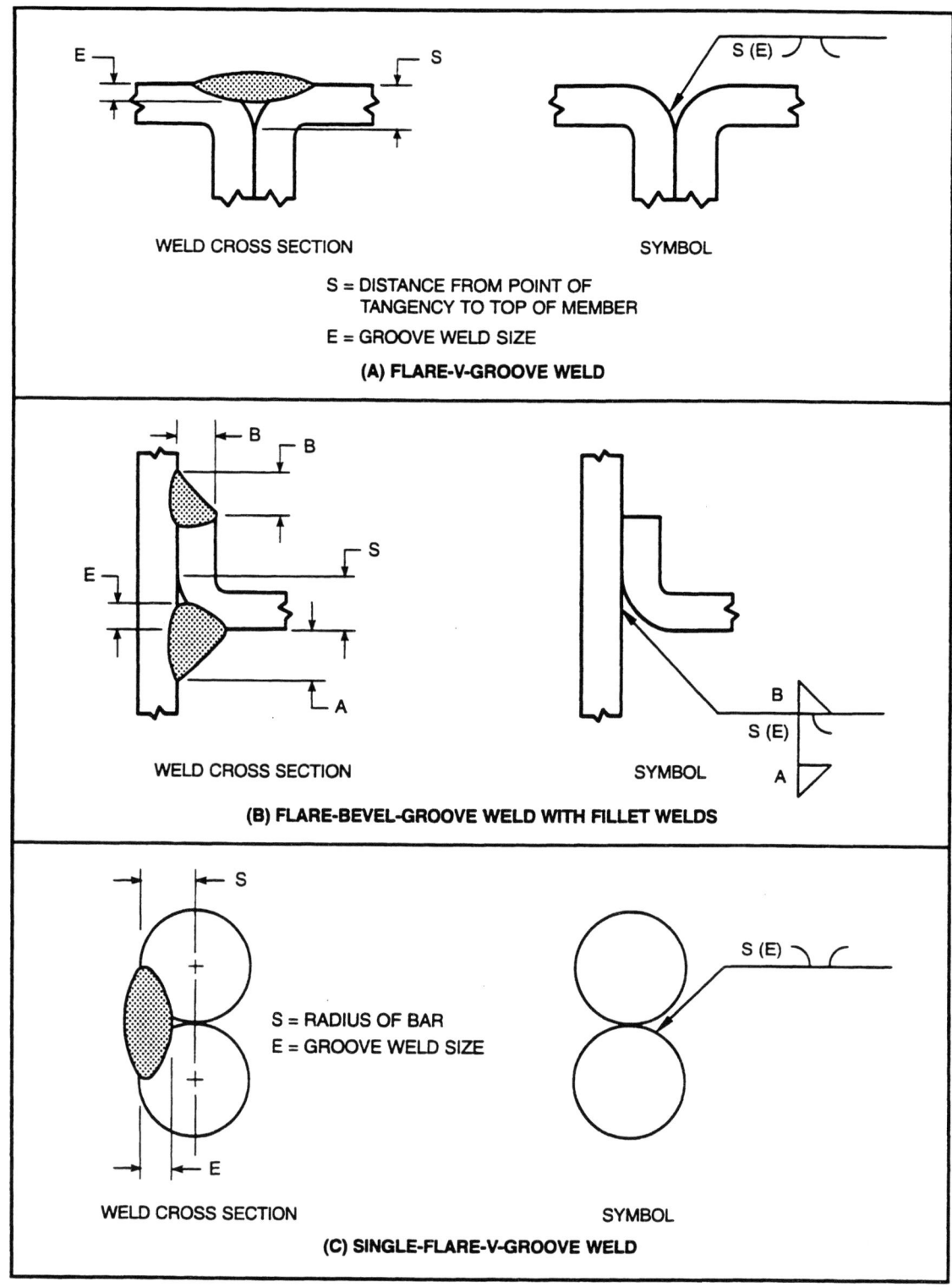

Figure 20—Applications of Flare-Bevel and Flare-V-Groove Weld Symbols

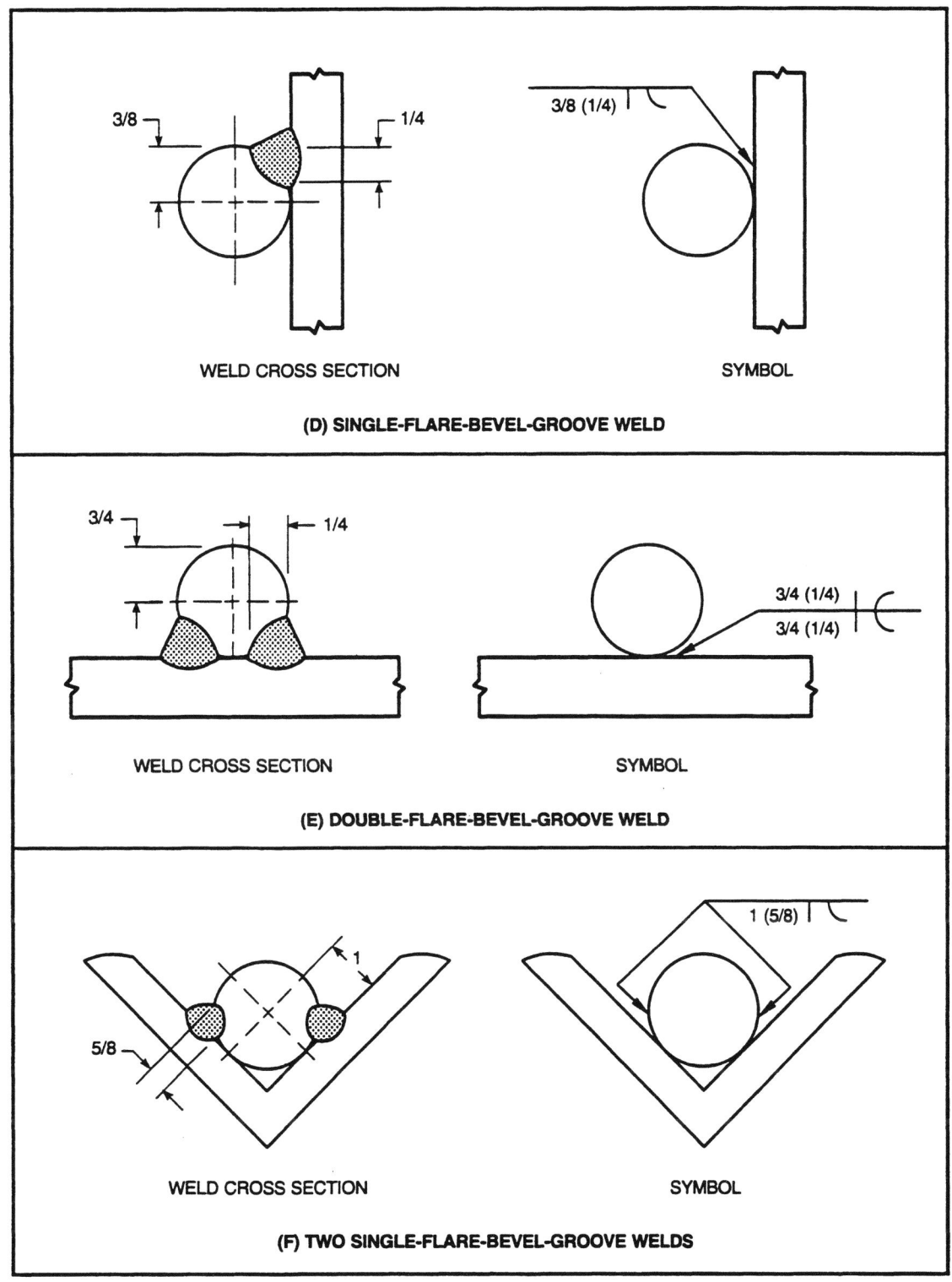

Figure 20 (Continued)—Applications of Flare-Bevel and Flare-V-Groove Weld Symbols

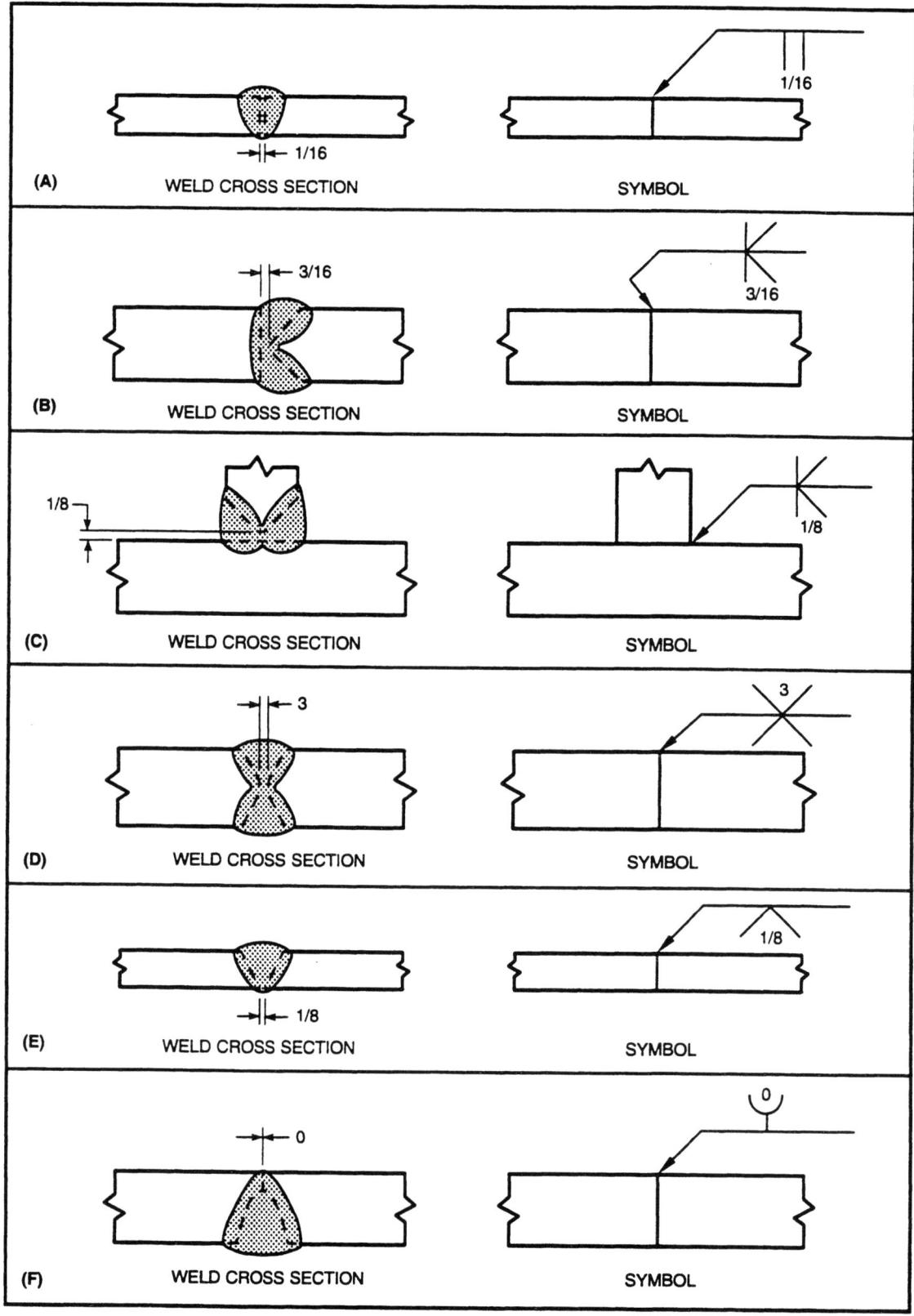

Figure 21—Specification of Root Opening of Groove Welds

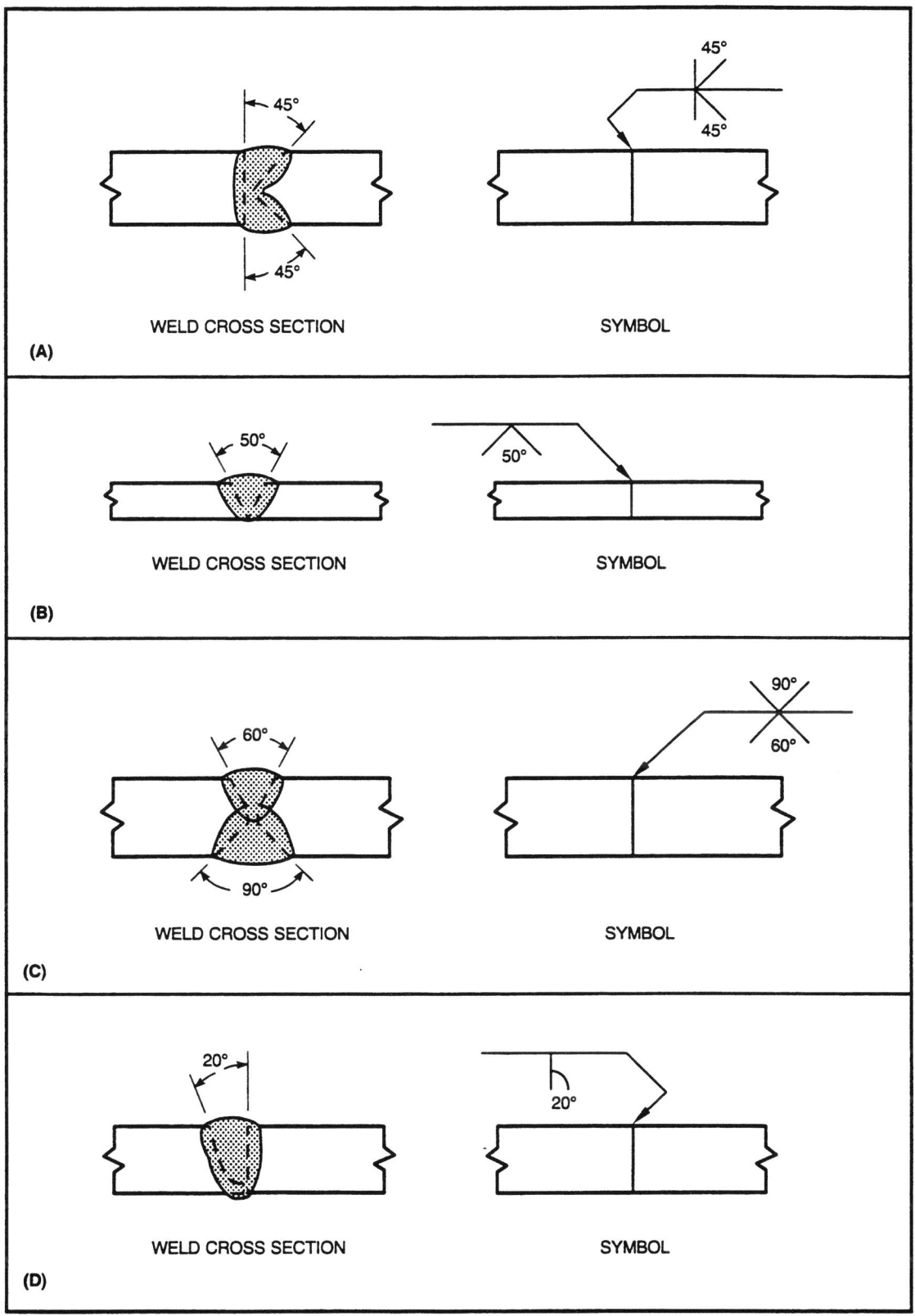

Figure 22—Specification of Groove Angle of Groove Welds

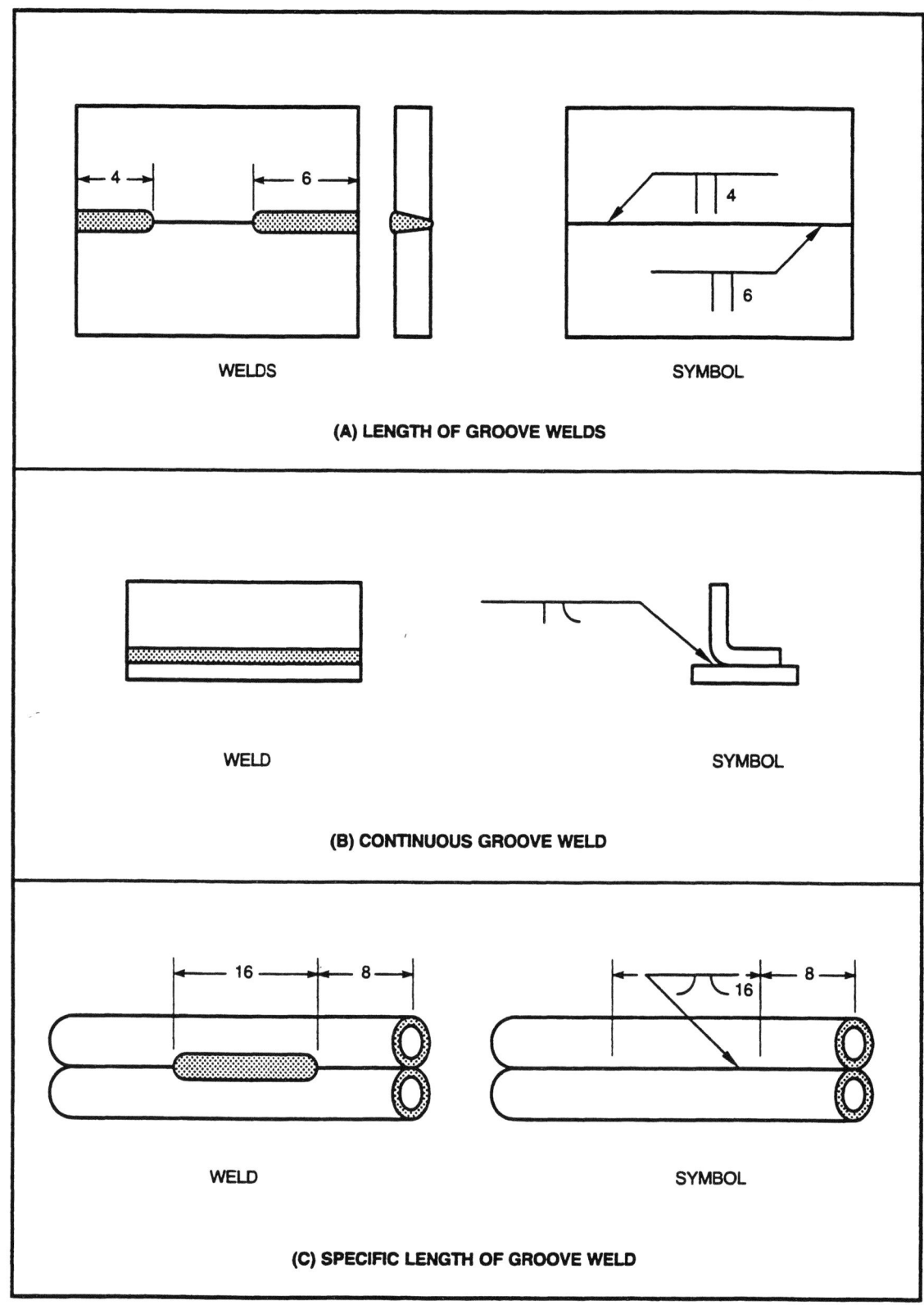

Figure 23—Specification of Length of Groove Welds

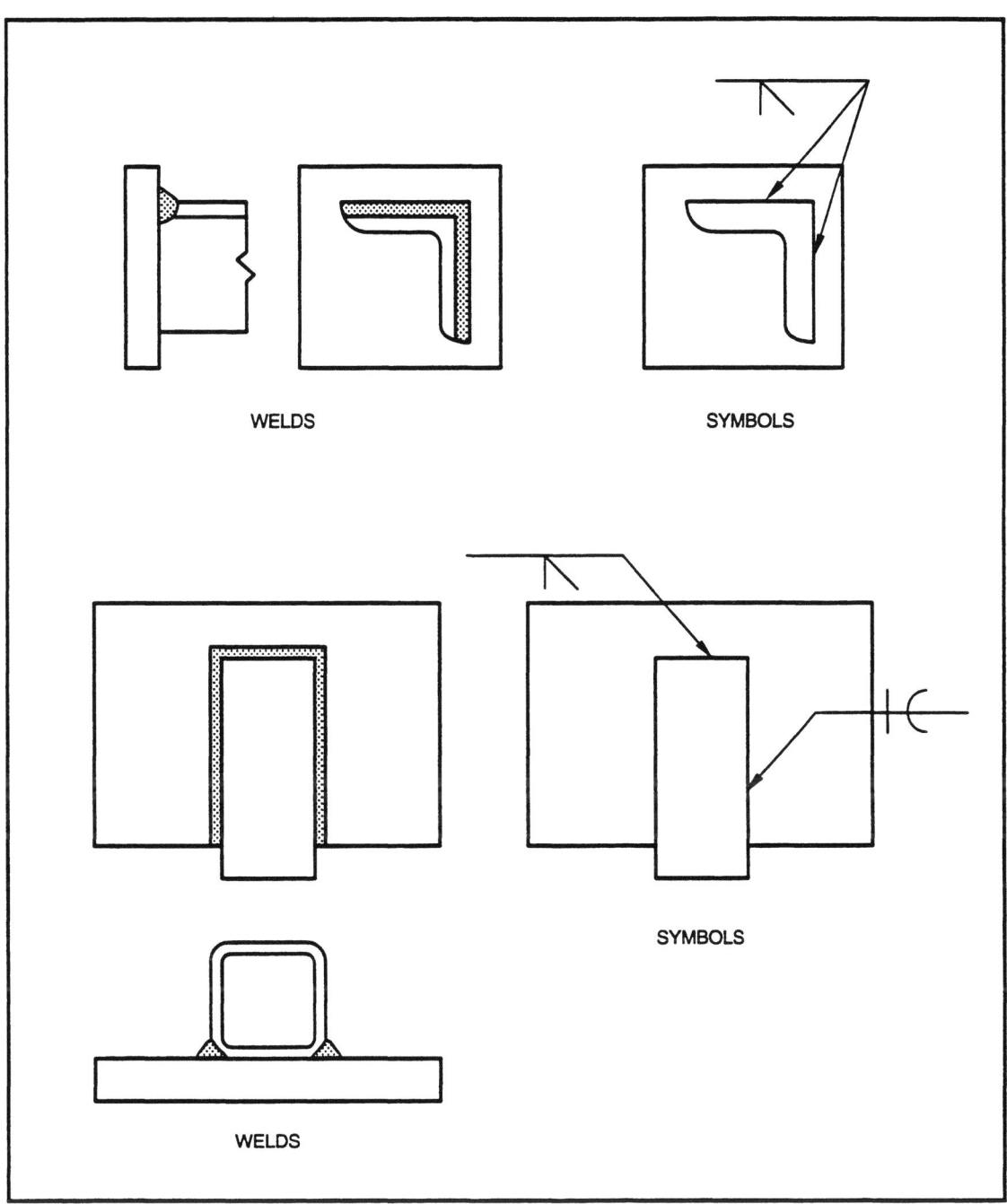

Figure 24—Specification of Extent of Welding for Groove Welds

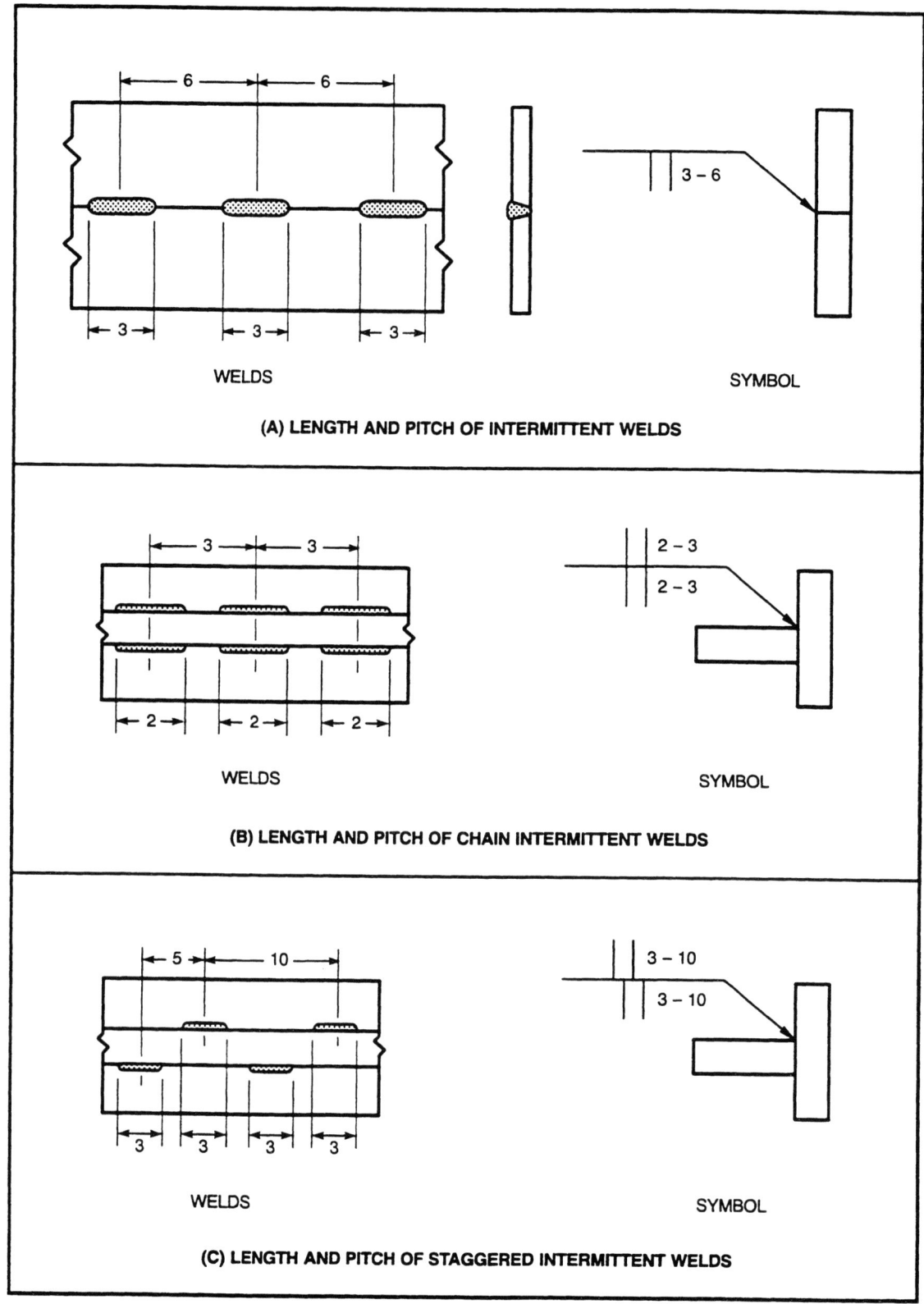

Figure 25—Applications of Intermittent Welds

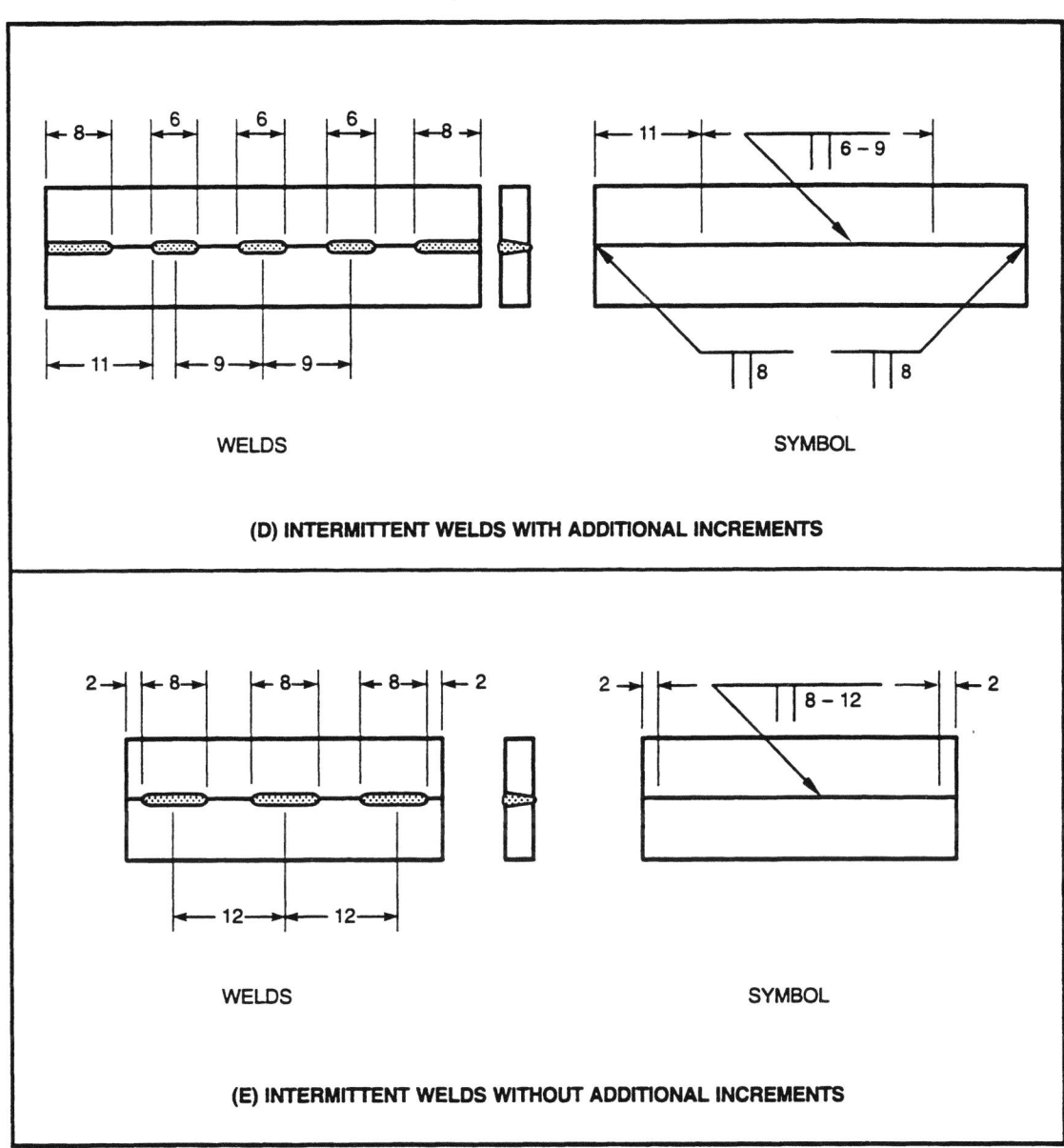

Figure 25 (Continued)—Applications of Intermittent Welds

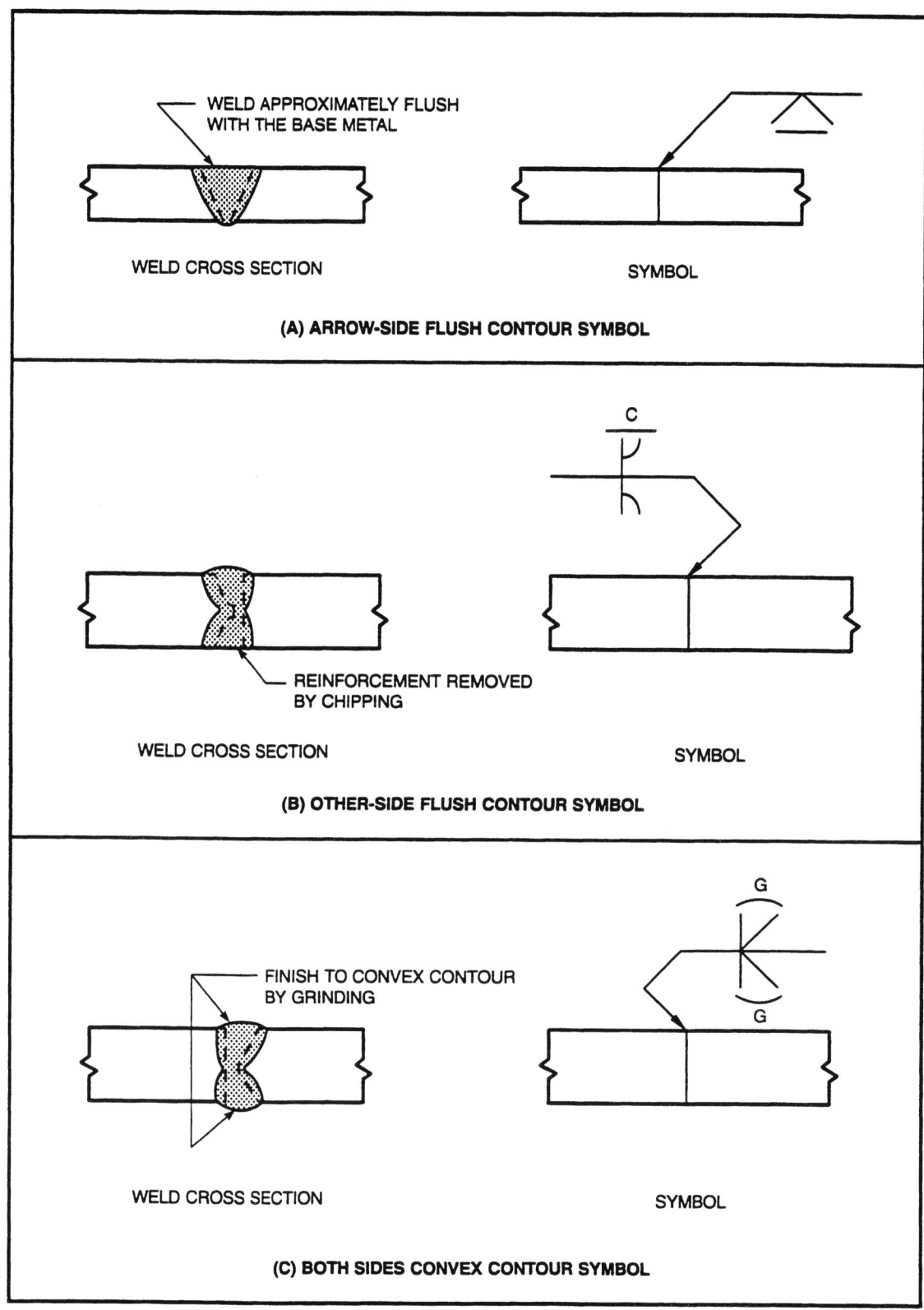

Figure 26—Applications of Flush and Convex Contour Symbols

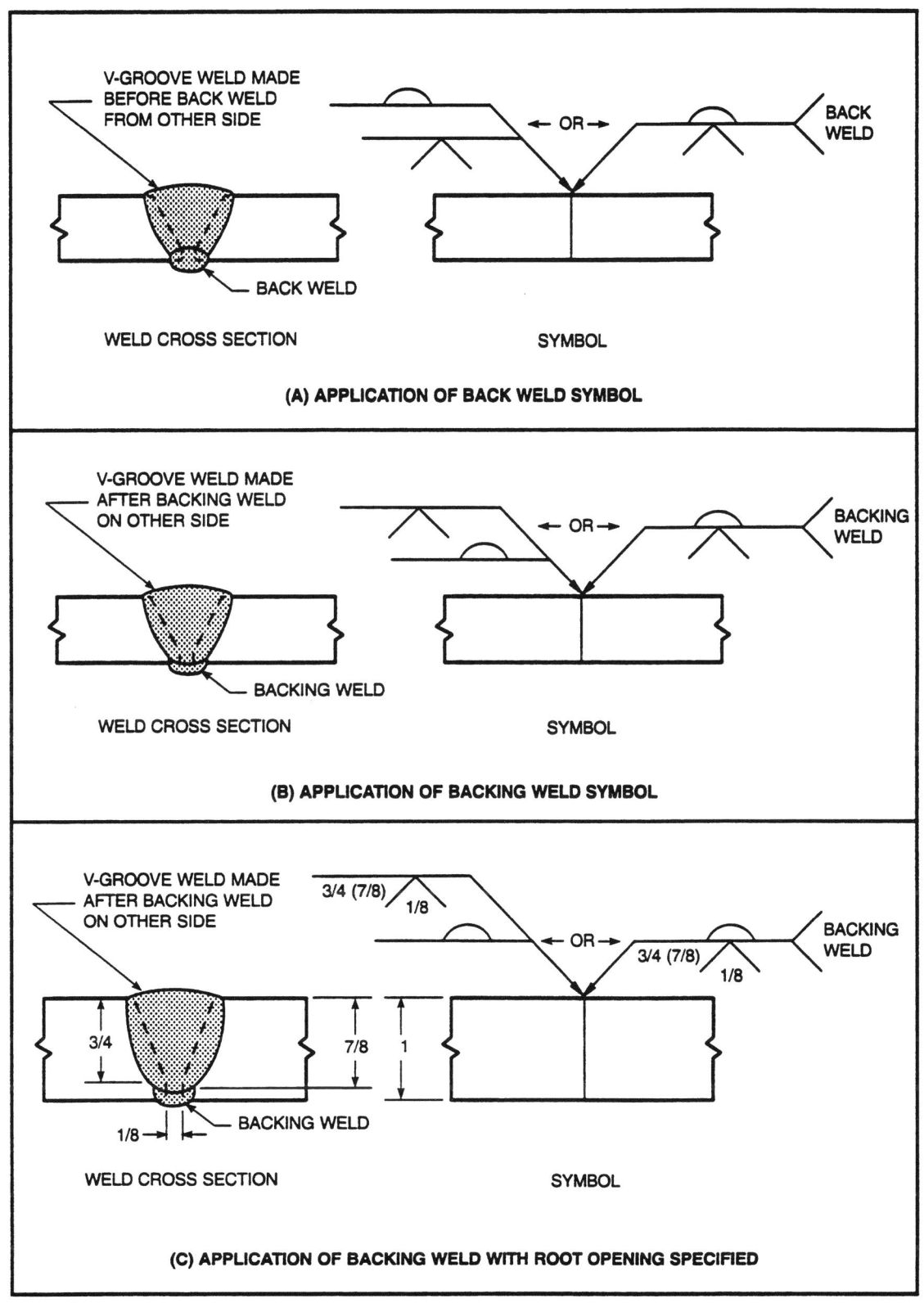

Figure 27—Applications of Back or Backing Weld Symbol

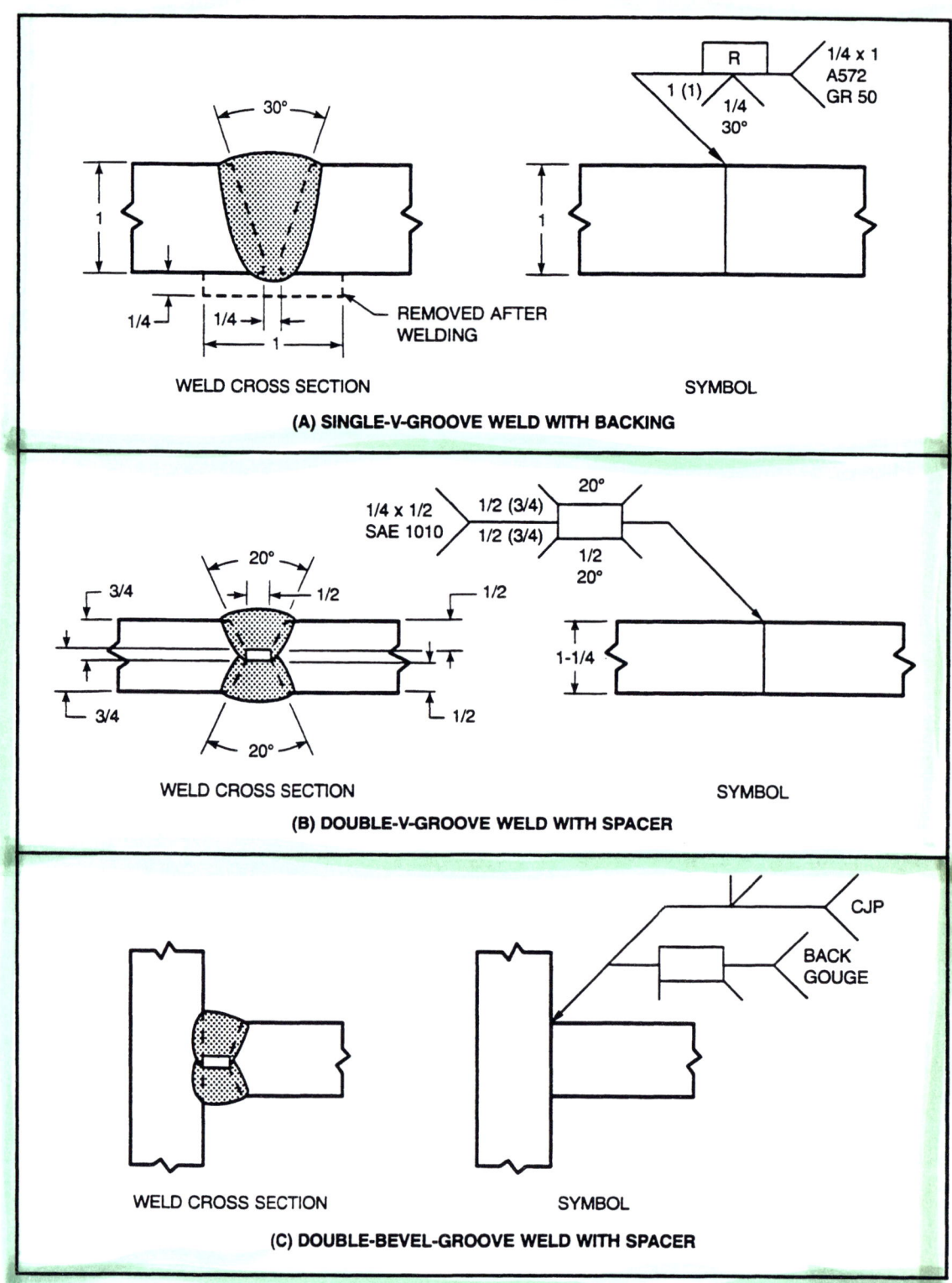

Figure 28—Joints with Backing or Spacers

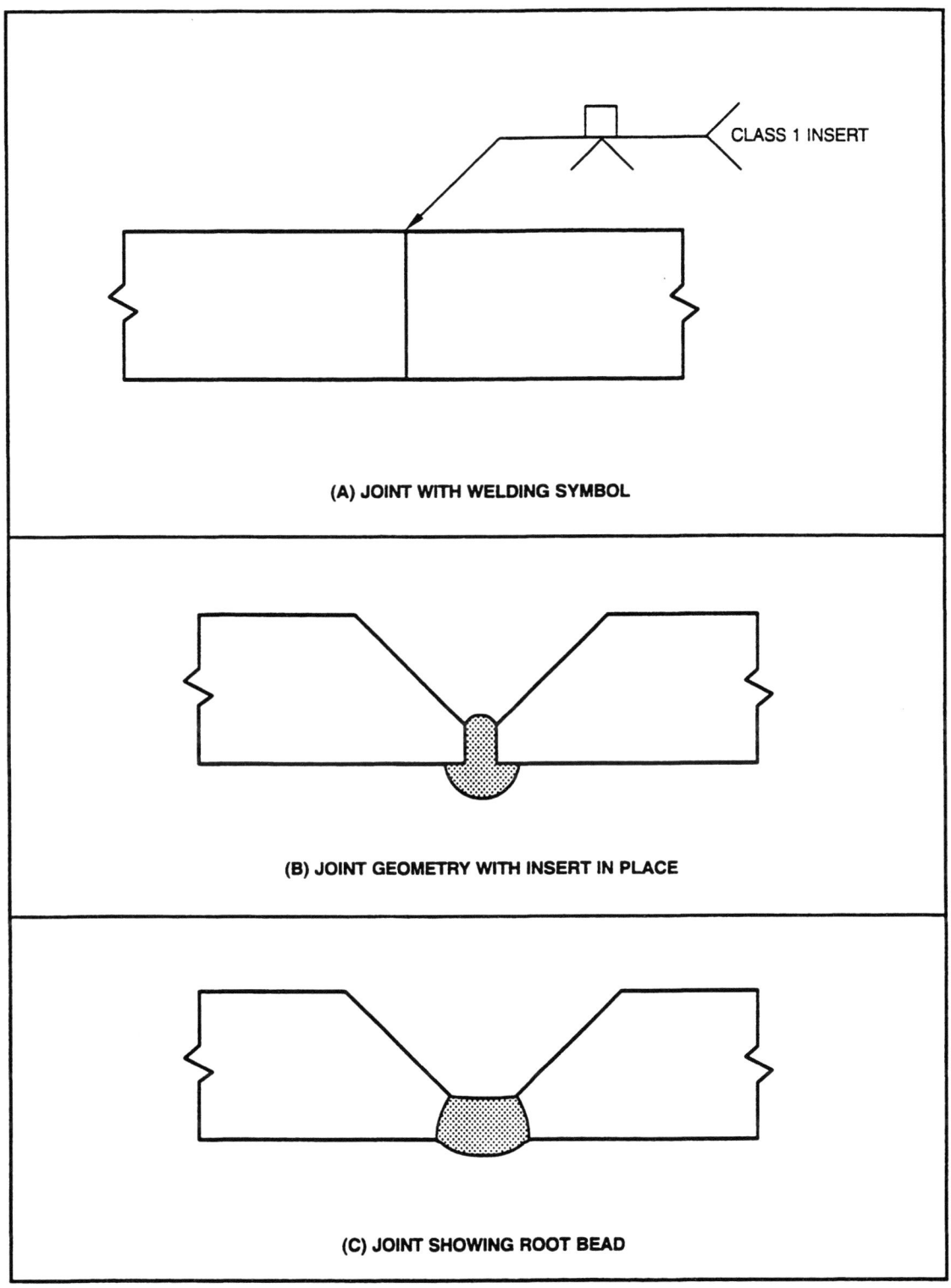

Figure 29—Application of the Consumable Insert Symbol

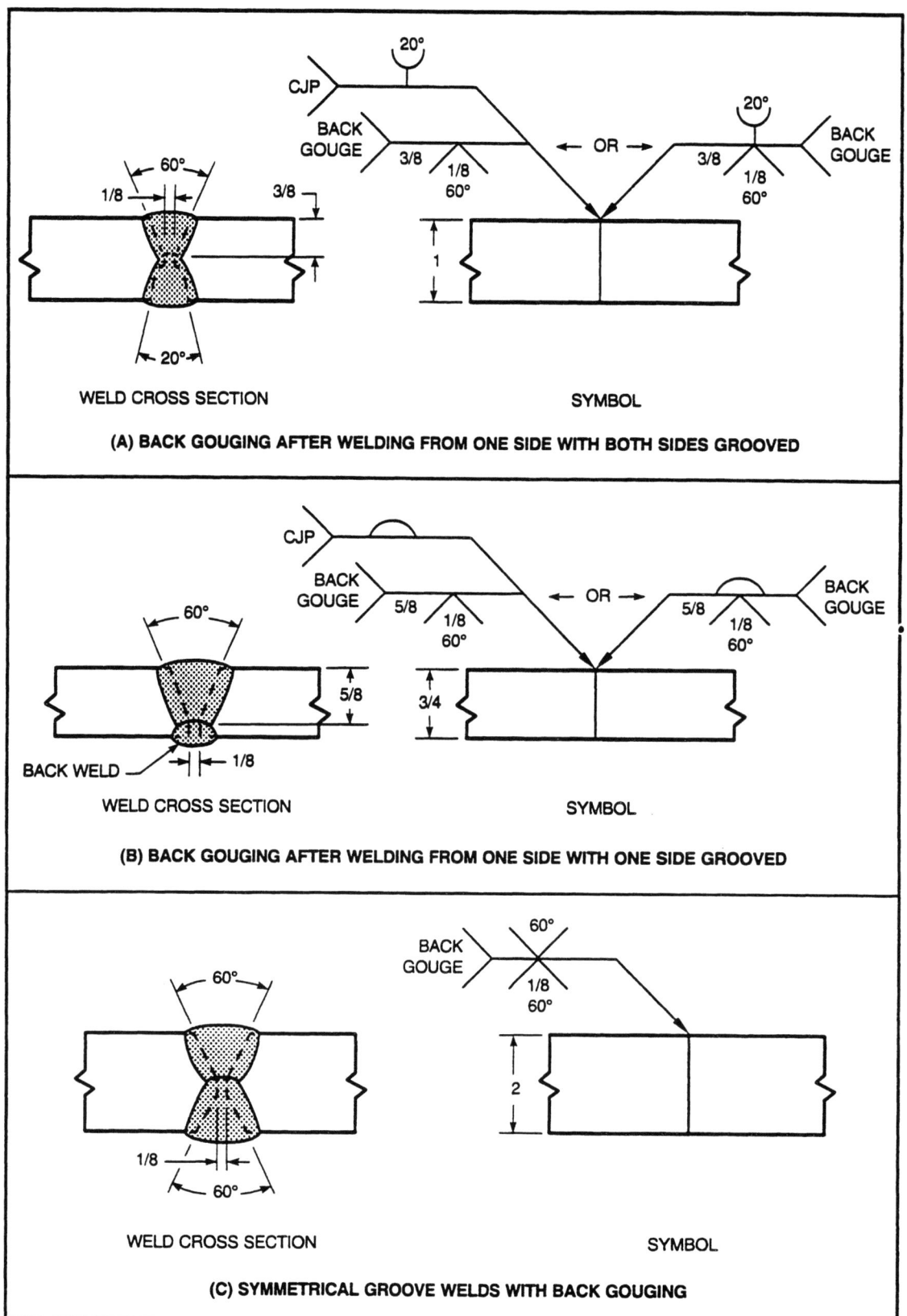

Figure 30—Groove Welds with Backgouging

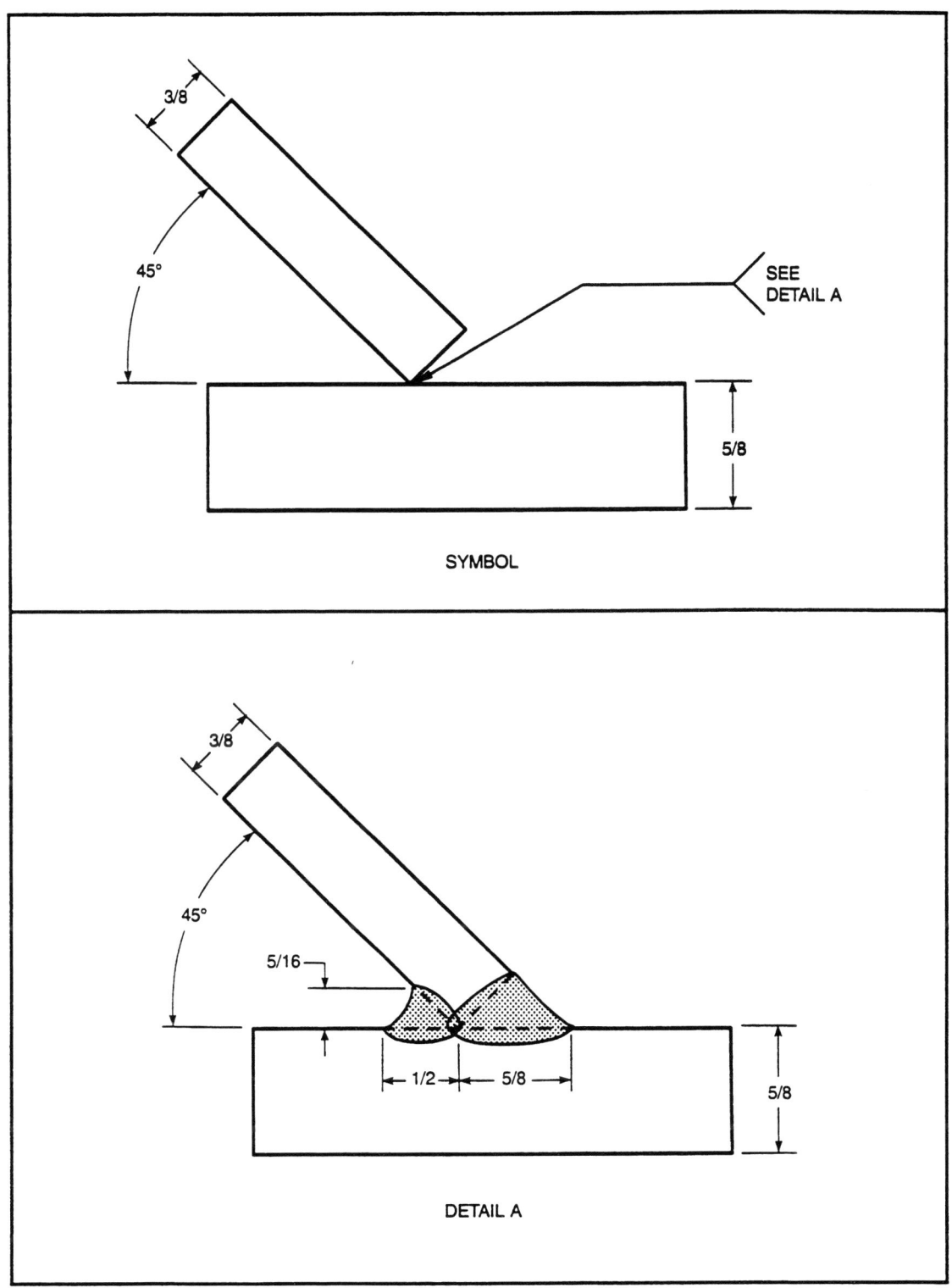

Figure 31—Skewed Joint

5. Fillet Welds

5.1 General

5.1.1 Dimension Location. Dimensions of fillet welds shall be shown on the same side of the reference line as the weld symbol (see Figures 32–34).

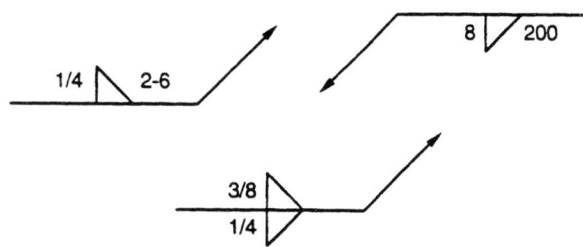

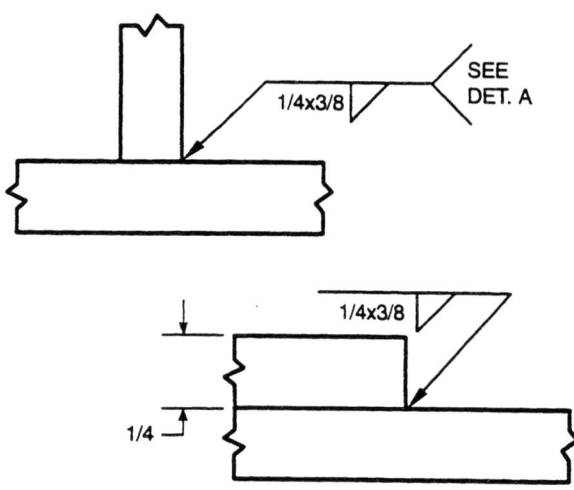

5.3 Length of Fillet Welds

5.3.1 Location. The length of a fillet weld, when indicated on the welding symbol, shall be specified to the right of the weld symbol [see Figure 32(F)].

5.1.2 Double Fillet Welds. The dimensions of fillet welds on both sides of a joint shall be specified whether the dimensions are identical or different [see Figures 32(B) and (C) and Figures 33(B) and (C)].

5.1.3 Drawing Notes. Dimensions of fillet welds covered by drawing notes need not be repeated on the welding symbols in accordance with 3.11.6.

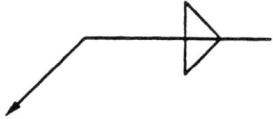

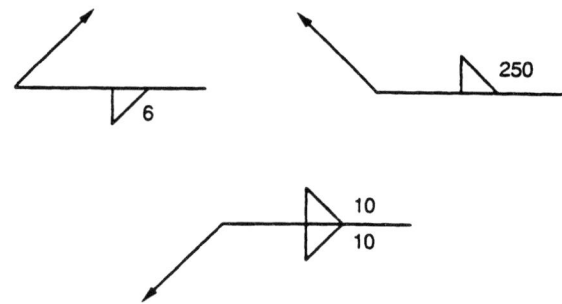

5.3.1.1 Full Length. When a fillet weld extends for the full length of the joint, no length dimension need be specified on the welding symbol [see Figure 32(A), (B), (C), (D), and (E)].

5.2 Size of Fillet Welds

5.2.1 Location. The fillet weld size shall be specified to the left of the weld symbol (see Figure 32).

5.3.1.2 Specific Lengths. Specific lengths of fillet welds, and their location, may be specified by symbols in conjunction with dimension lines [see Figures 8(C) and 32(F)].

5.3.1.3 Hatching. Hatching may be used to graphically depict fillet welds (see 4.4.1.3).

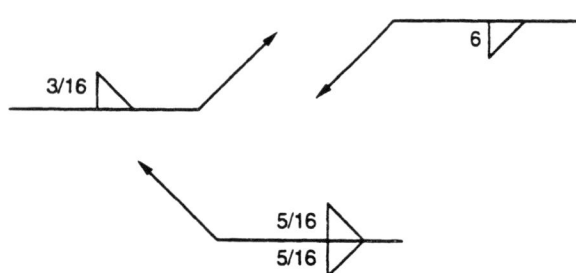

5.2.2 Unequal Legs. The size of a fillet weld with unequal legs shall be specified to the left of the weld symbol as shown below. Weld orientation is not specified by the symbol and shall be shown on the drawing to ensure clarity [see Figure 32(D)].

5.3.2 Changes in Direction of Welding. Symbols for fillet welds involving changes in the direction of welding shall be in accordance with 3.9.2 [see Figure 9(A)].

5.4 Intermittent Fillet Welds

5.4.1 Pitch. The pitch of intermittent fillet welds shall be the distance between the centers of adjacent weld segments on one side of the joint [see Figure 25(C)].

5.4.2 Pitch Dimension Location. The pitch of intermittent fillet welds shall be specified to the right of the length dimension following a hyphen (see Figure 33).

5.4.3 Chain Intermittent Fillet Welds. Dimensions of chain intermittent fillet welds shall be specified on both sides of the reference line. The segments of chain intermittent fillet welds shall be opposite one another across the joint [see Figure 33(B)].

5.4.4 Staggered Intermittent Fillet Welds. Dimensions of staggered intermittent fillet welds shall be specified on both sides of the reference line, and the fillet weld symbols shall be offset on opposite sides of the reference line as shown below. The segments of staggered intermittent fillet welds shall be symmetrically spaced on both sides of the joint as shown in Figure 33(C).

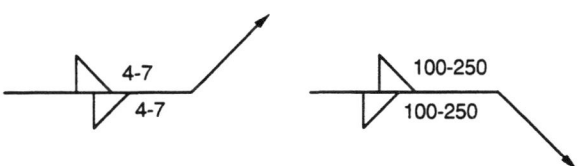

5.4.5 Extent of Welding. In the case of intermittent fillet welds, additional weld lengths which are intended at the ends of the joint shall be specified by separate welding symbols and dimensioned on the drawing [see Figure 33(D)]. When no weld lengths are intended at the ends of the joint, the unwelded lengths should not exceed the clear distance between weld segments and be so dimensioned on the drawing [see Figure 33(E)].

5.4.6 Location of Intermittent Welds. When the location of intermittent welds is not obvious, such as on a circular weld joint, it will be necessary to provide specific segment locations by dimension lines (see 4.4.1.2 and 5.3.1.2) or by hatching (see 4.4.1.3 and 5.3.1.3).

5.5 Fillet Welds in Holes and Slots. Fillet welds in holes and slots shall be specified by the use of fillet weld symbols [see Figure 34(A)].

5.6 Contours and Finishing of Fillet Welds

5.6.1 Contours Obtained by Welding. Fillet welds that are to be welded with approximately flat, convex or concave faces without postweld finishing shall be specified by adding the flat, convex, or concave contour symbol to the welding symbol as follows (see 3.12).

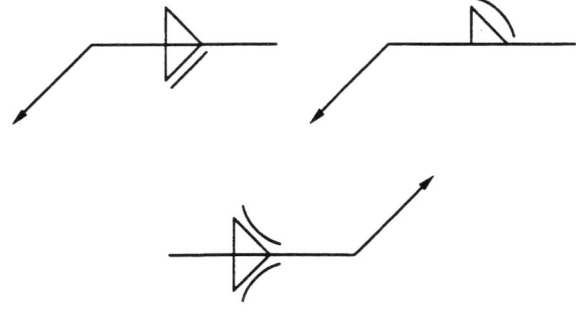

5.6.2 Contours Obtained by Postweld Finishing. Fillet welds that are to be finished approximately flat, convex, or concave by postweld finishing shall be specified by adding both the appropriate contour and finishing symbols to the welding symbol as follows (see 3.13).

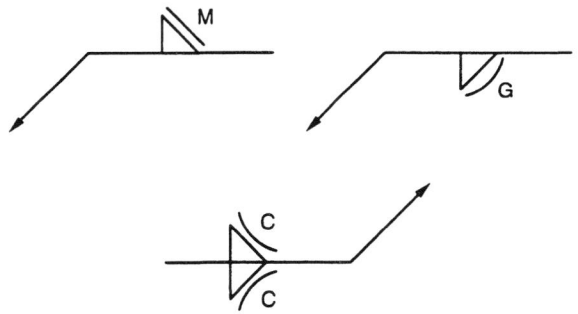

5.7 Skewed Joints. When the angle between the fusion faces is such that the identification of the weld type and, hence, proper weld symbol may be in question, the detail of the desired joint and weld configuration shall be shown on the drawing [see 4.13 and Figure 31].

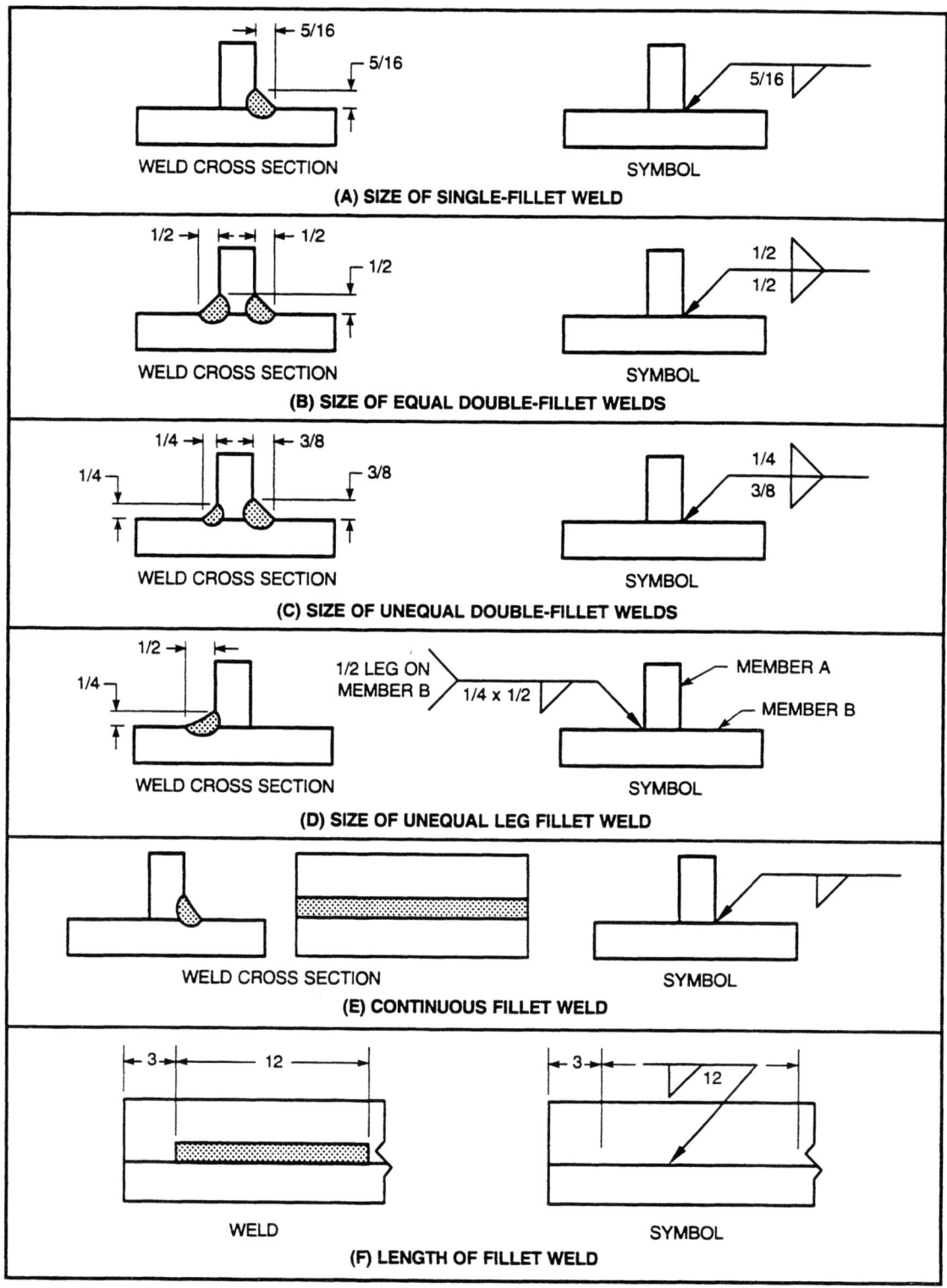

Figure 32—Specification of Size and Length of Fillet Welds

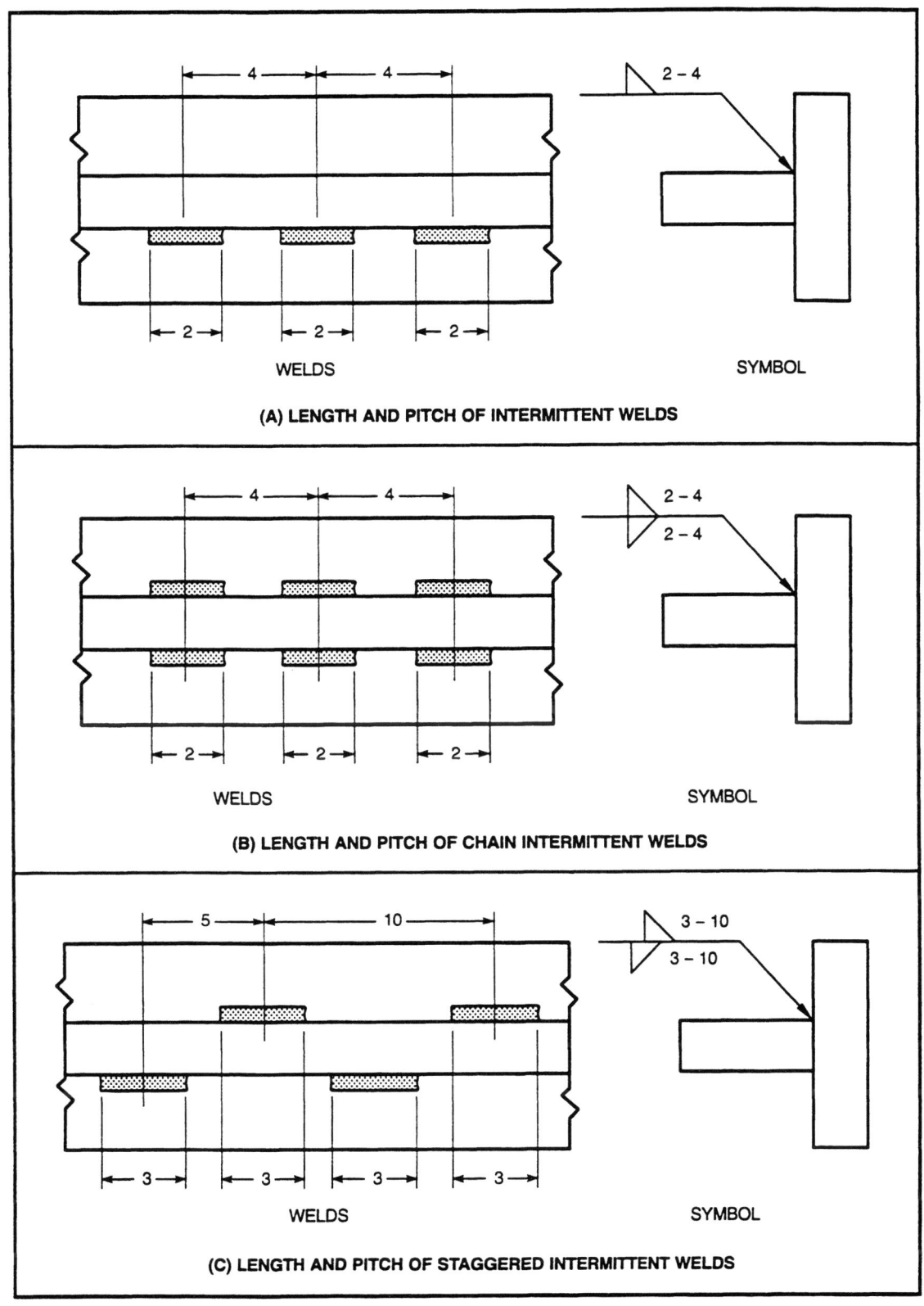

Figure 33—Applications of Intermittent Fillet Weld Symbols

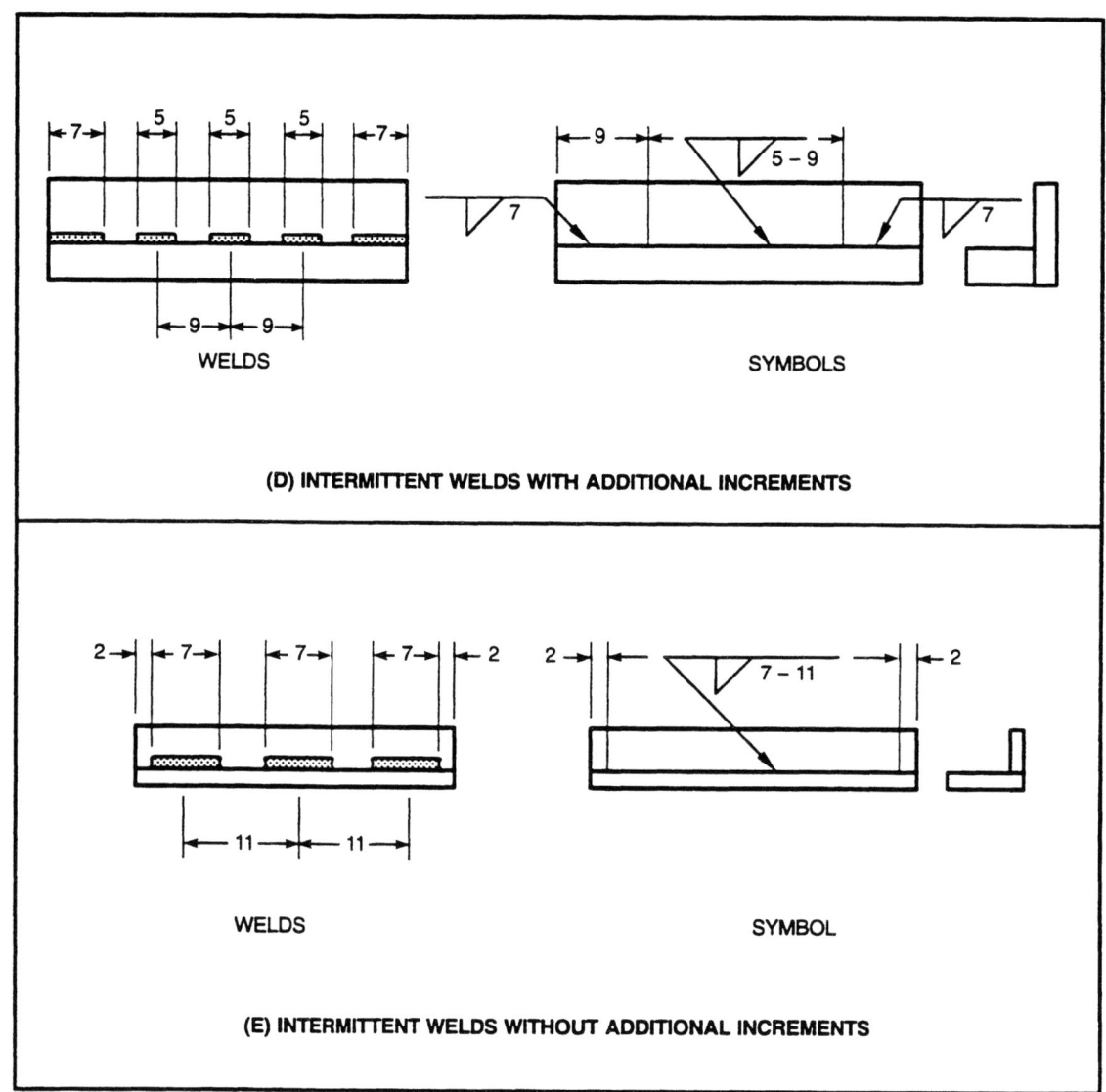

Figure 33 (Continued)—Applications of Intermittent Fillet Weld Symbols

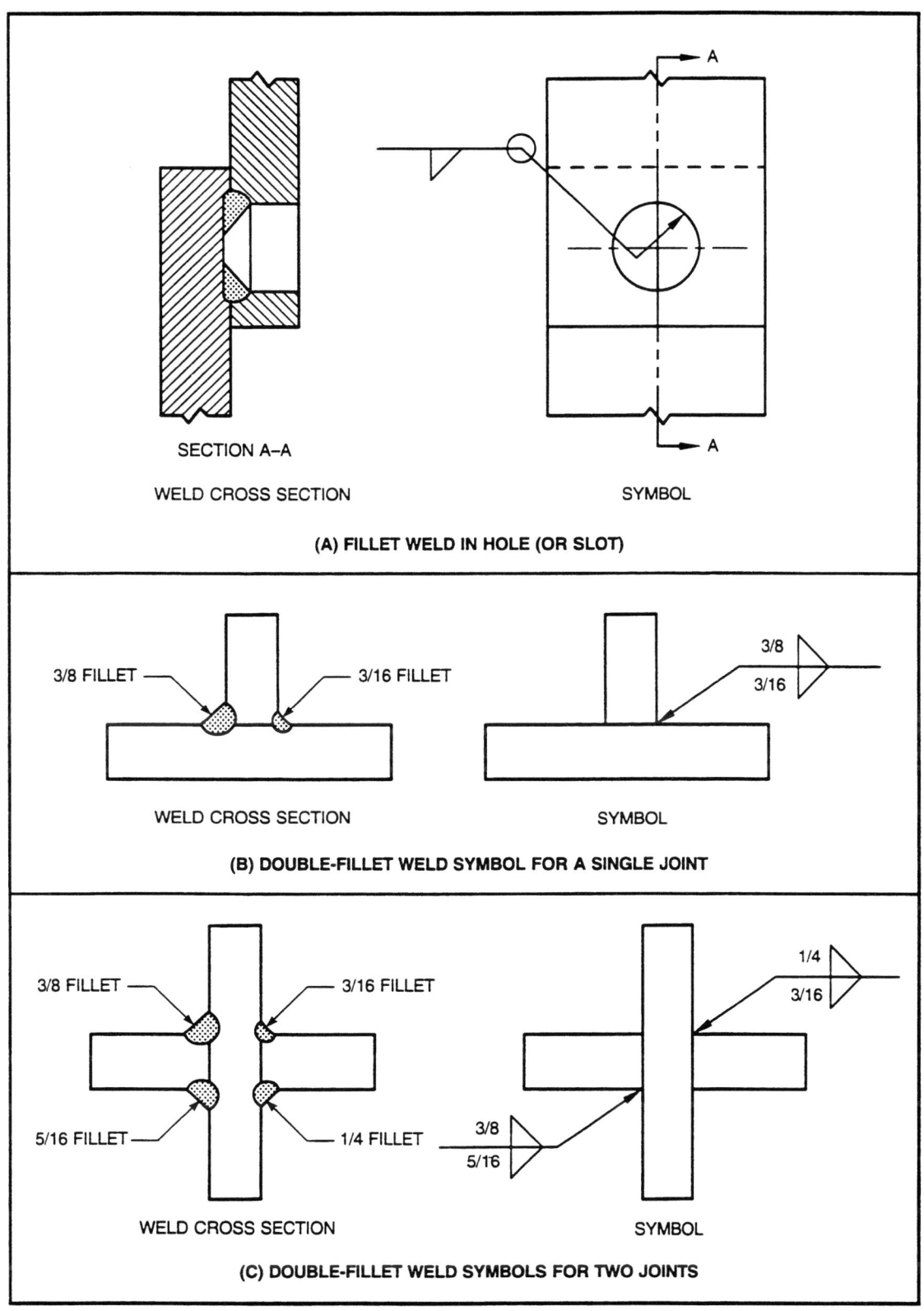

Figure 34—Applications of Fillet Weld Symbol

6. Plug Welds

6.1 General

6.1.1 Arrow-Side Holes. Holes in the arrow-side member of a joint to be plug welded shall be specified by placing the plug weld symbol below the reference line [see Figure 35(A)].

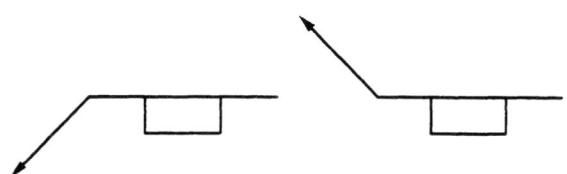

6.1.2 Other-Side Holes. Holes in the other-side member of a joint to be plug welded shall be specified by placing the plug weld symbol above the reference line [see Figure 35(B)].

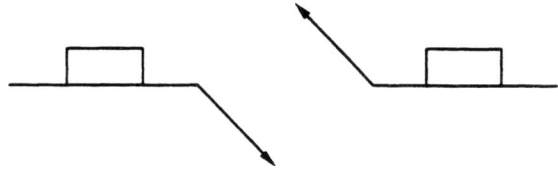

6.1.3 Dimensions. Dimensions of plug welds shall be specified on the same side of the reference line as the weld symbol (see Figure 36).

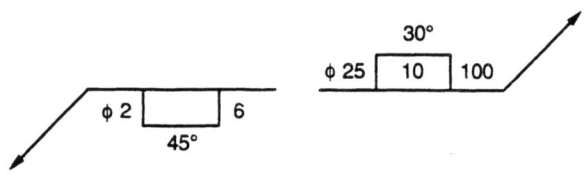

6.1.4 Fillets in Holes. The plug weld symbol shall not be used to designate fillet welds in holes (see 5.5).

6.2 Plug Weld Size. The plug weld size shall be specified to the left of the plug weld symbol and shall be preceded by the diameter symbol, φ, shown as follows [see Figure 36(A), (E), (F), and (G)]. Plug weld size is the diameter of the hole at the faying surface.

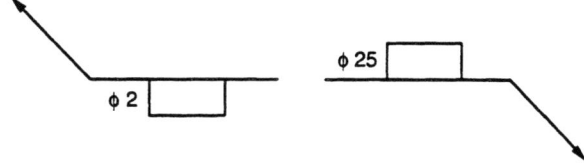

6.3 Angle of Countersink. The included angle of countersink of plug welds shall be located on the same side of the reference line and above or below the plug weld symbol as appropriate [see Figure 36(B) and (E)].

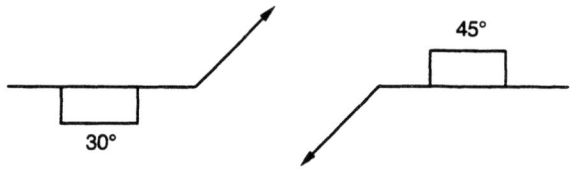

6.4 Depth of Filling. When the depth of filling is less than complete, it shall be specified inside the plug weld symbol [see Figure 36(C) and (E)]. The omission of a depth dimension shall specify complete filling [see Figure 36(A), (B), (D), (F), and (G)].

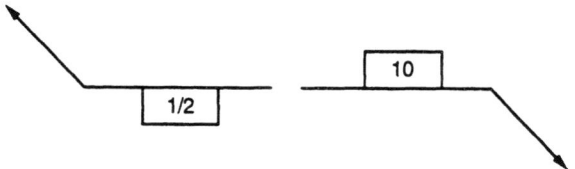

6.5 Spacing of Plug Welds. The pitch (center-to-center distance) of plug welds in a straight line shall be specified to the right of the plug weld symbol [see Figure 36(D) and (E)]. The spacing of plug welds in any configuration other than a straight line shall be dimensioned on the drawing.

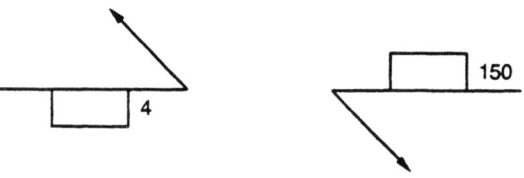

6.6 Number of Plug Welds. When a definite number of plug welds is desired in a joint, the number shall be specified in parentheses on the same side of the reference line as the weld symbol. The number shall be either above or below the weld symbol, as appropriate [see Figure 36(D) and (E)]. When the welding symbol also includes the angle of countersink, the number of plug welds shall be placed either above or below the angle of countersink as appropriate [see Figure 36(E)].

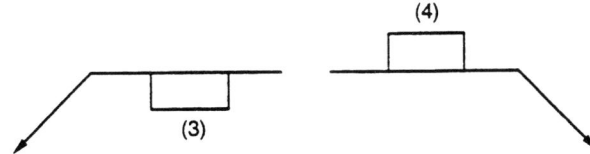

6.7 Contours and Finishing of Plug Welds

6.7.1 Contours Obtained by Welding. Plug welds that are to be welded with approximately flush or convex faces without postweld finishing shall be specified by adding the flush or convex contour symbol to the welding symbol (see 3.12).

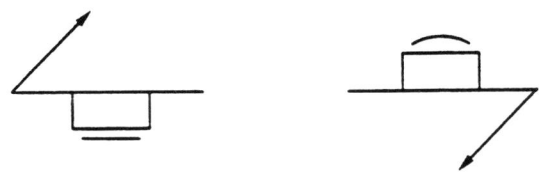

6.7.2 Contours Obtained by Postweld Finishing. Plug welds whose faces are to be finished approximately flush or convex by postweld finishing shall be specified by adding both the appropriate contour and finishing symbols to the welding symbol (see 3.13). Welds that require a flat but not flush surface require an explanatory note in the tail of the welding symbol.

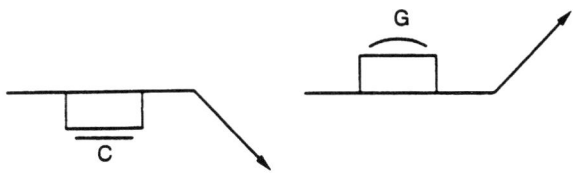

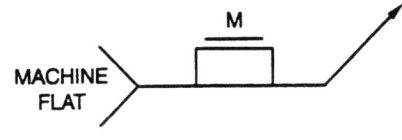

6.8 Joints Involving Three or More Members. Plug welding symbols may be used to specify welding two or more members to another member. A section view of the joint shall be provided to clarify which members require preparation [see Figure 36(F) and (G)].

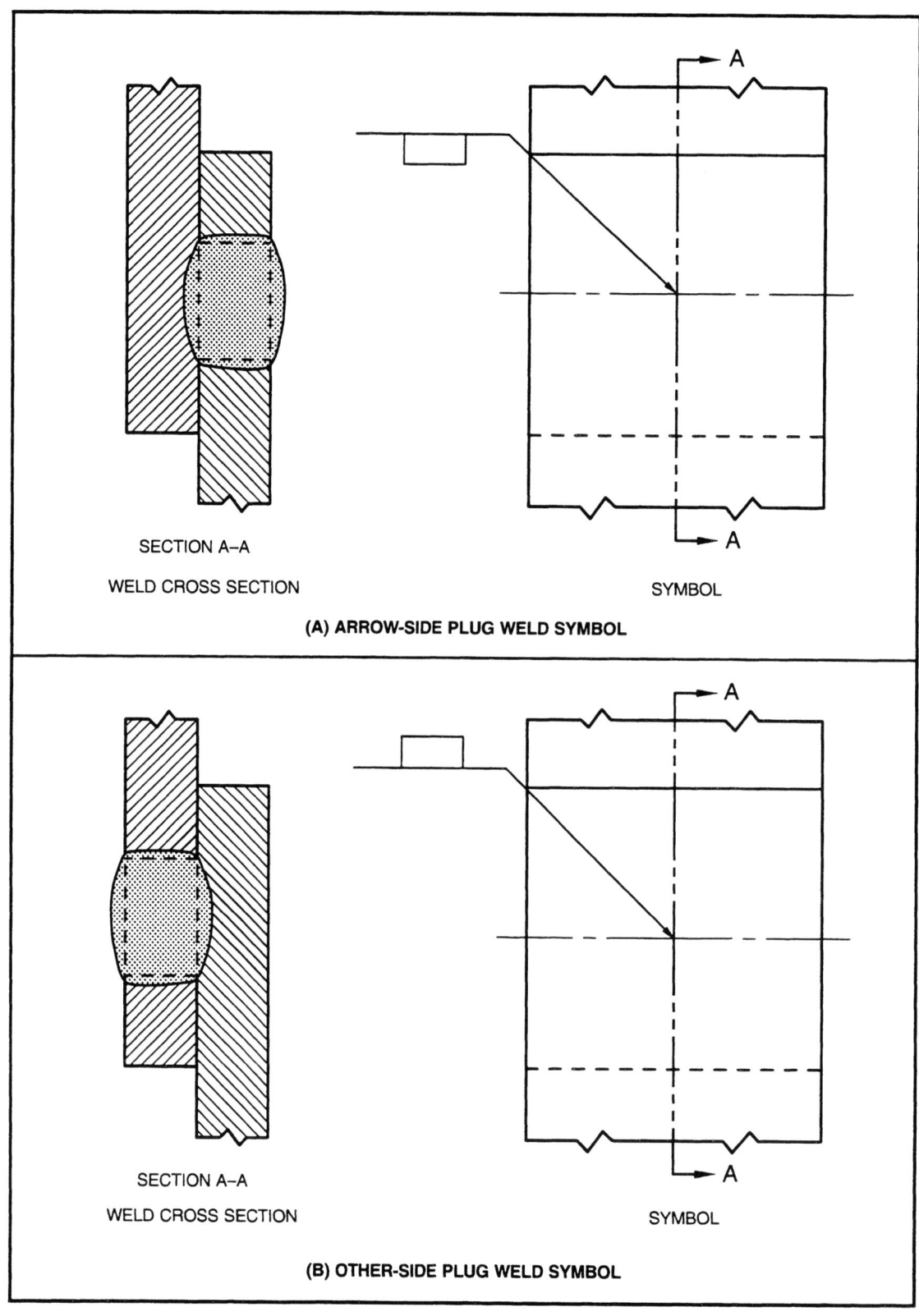

Figure 35—Applications of Plug Weld Symbol

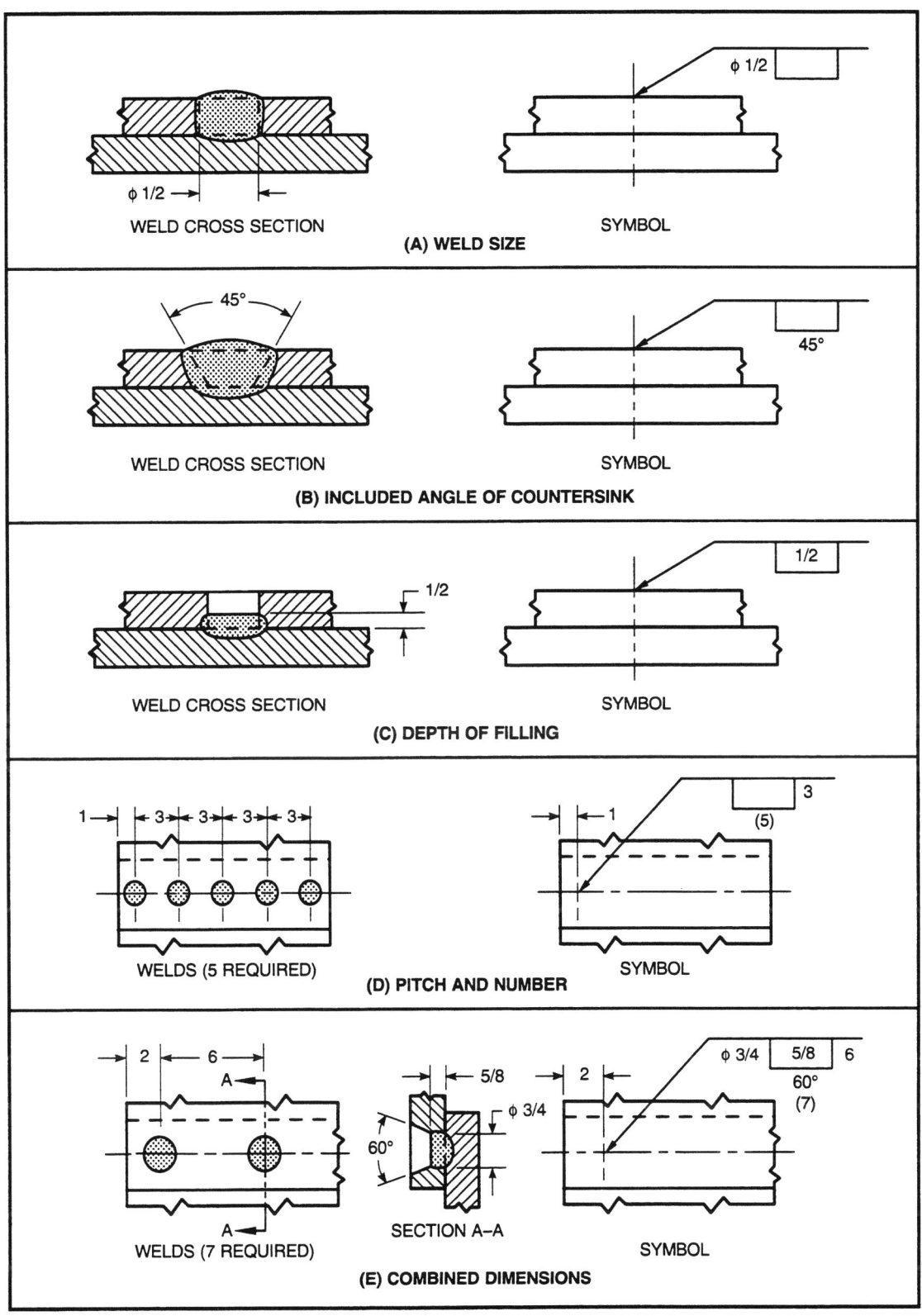

Figure 36—Applications of Information to Plug Weld Symbols

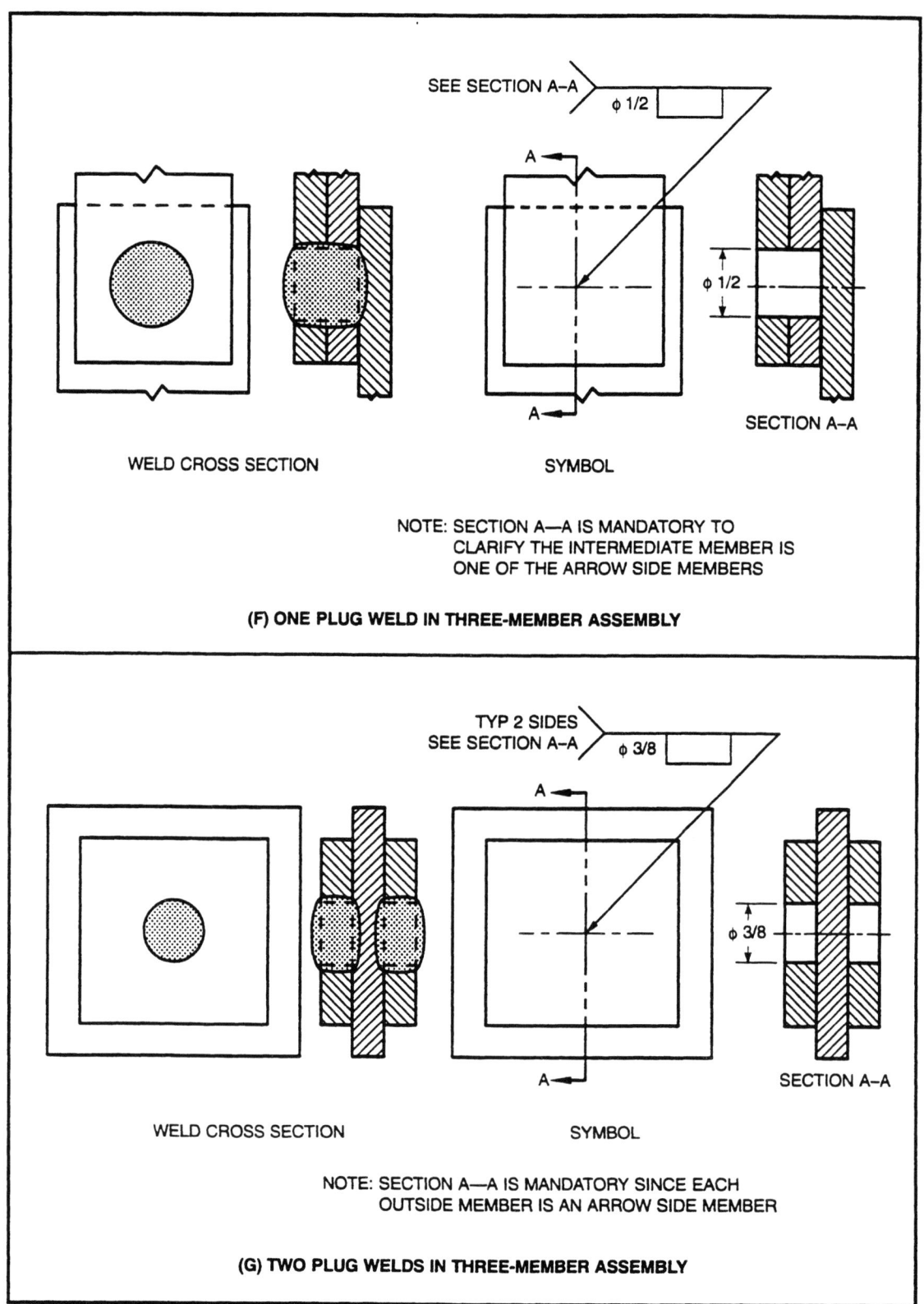

Figure 36 (Continued)—Applications of Information to Plug Weld Symbols

7. Slot Welds

7.1 General

7.1.1 Arrow-Side Slots. Slots in the arrow-side member of a joint to be slot welded shall be specified by placing the slot weld symbol below the reference line [see Figure 37(A)].

7.1.2 Other-Side Slots. Slots in the other-side member of a joint to be slot welded shall be specified by placing the slot weld symbol above the reference line [see Figure 37(B)].

7.1.3 Dimensions. Dimensions of slot welds shall be specified on the same side of the reference line as the weld symbol (see Figure 38).

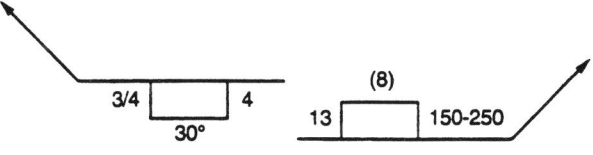

7.1.4 Fillets in Slots. The slot weld symbol shall not be used to specify fillet welds in slots (see 5.5).

7.2 Width of Slot Welds.
The width of a slot weld shall be specified to the left of the weld symbol (see Figure 38). Slot weld width is the dimension of the slot, measured in the direction of the minor axis, at the faying surface.

7.3 Length of Slot Welds.
The length of slot welds shall be specified to the right of the weld symbol (see Figure 38). Slot weld length is the dimension of the slot, measured in the direction of the major axis, at the faying surface.

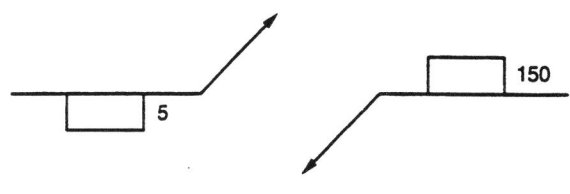

7.4 Angle of Countersink.
The countersink included angle of slot welds shall be specified either above or below the slot weld symbol as appropriate [see Figure 38(A)].

7.5 Depth of Filling.
Depth of filling less than complete shall be specified inside the slot weld symbol [see Figure 38(B)]. Omission of the depth dimension shall specify complete filling [see Figure 38(A)].

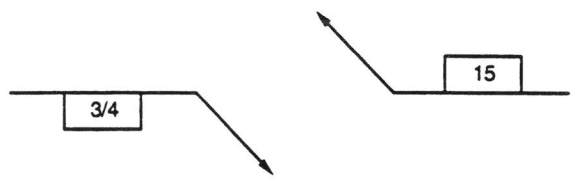

7.6 Spacing of Slot Welds.
The pitch (center-to-center distance) of slot welds in a straight line shall be specified to the right of the length dimension following a hyphen (see Figure 38).

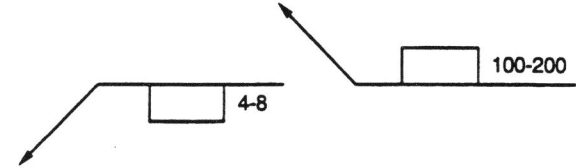

7.7 Number of Slot Welds. When a definite number of slot welds is desired in a joint, the number shall be specified in parentheses on the same side of the reference line as the weld symbol. The number shall be either above or below the weld symbol, as appropriate (see Figure 38). When the angle of countersink is also included in the welding symbol, the number of slot welds shall be placed above or below the angle of countersink as appropriate [see Figure 38(A)].

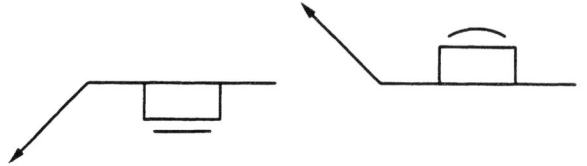

7.9.2 Contours Obtained by Postweld Finishing. Slot welds whose faces are to be finished approximately flush or convex by postweld finishing shall be specified by adding both the appropriate contour and finishing symbols to the welding symbol (see 3.13). Welds that require a flat but not flush surface require an explanatory note in the tail of the welding symbol.

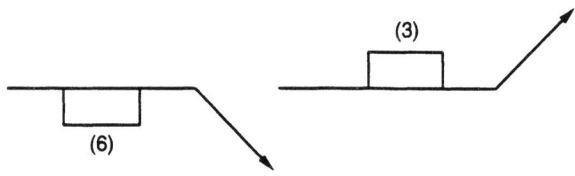

7.8 Location and Orientation of Slot Welds. The location and orientation of slot welds shall be specified on the drawing.

7.9 Contours and Finishing of Slot Welds

7.9.1 Contours Obtained by Welding. Slot welds that are to be welded with approximately flush or convex faces without postweld finishing shall be specified by adding the flush or convex contour symbol to the welding symbol (see 3.12).

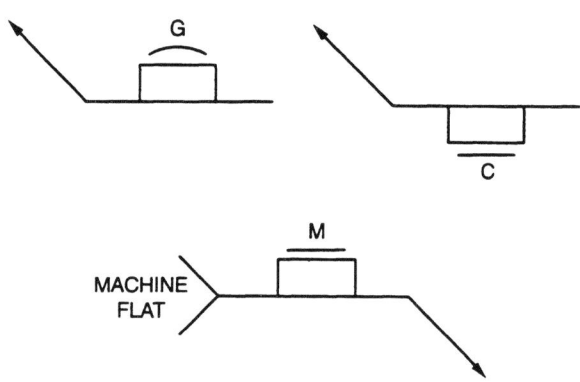

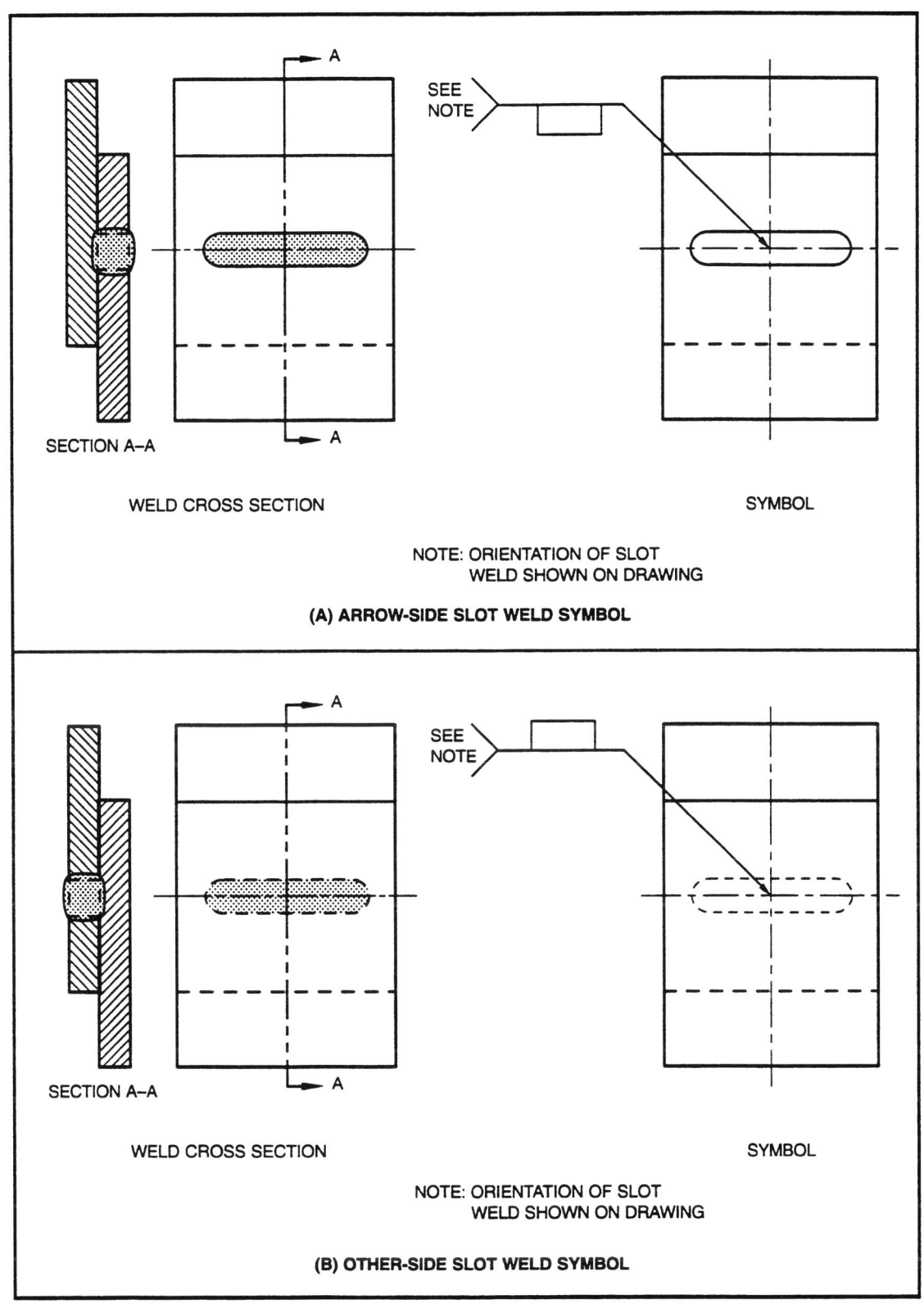

Figure 37—Applications of Slot Weld Symbol

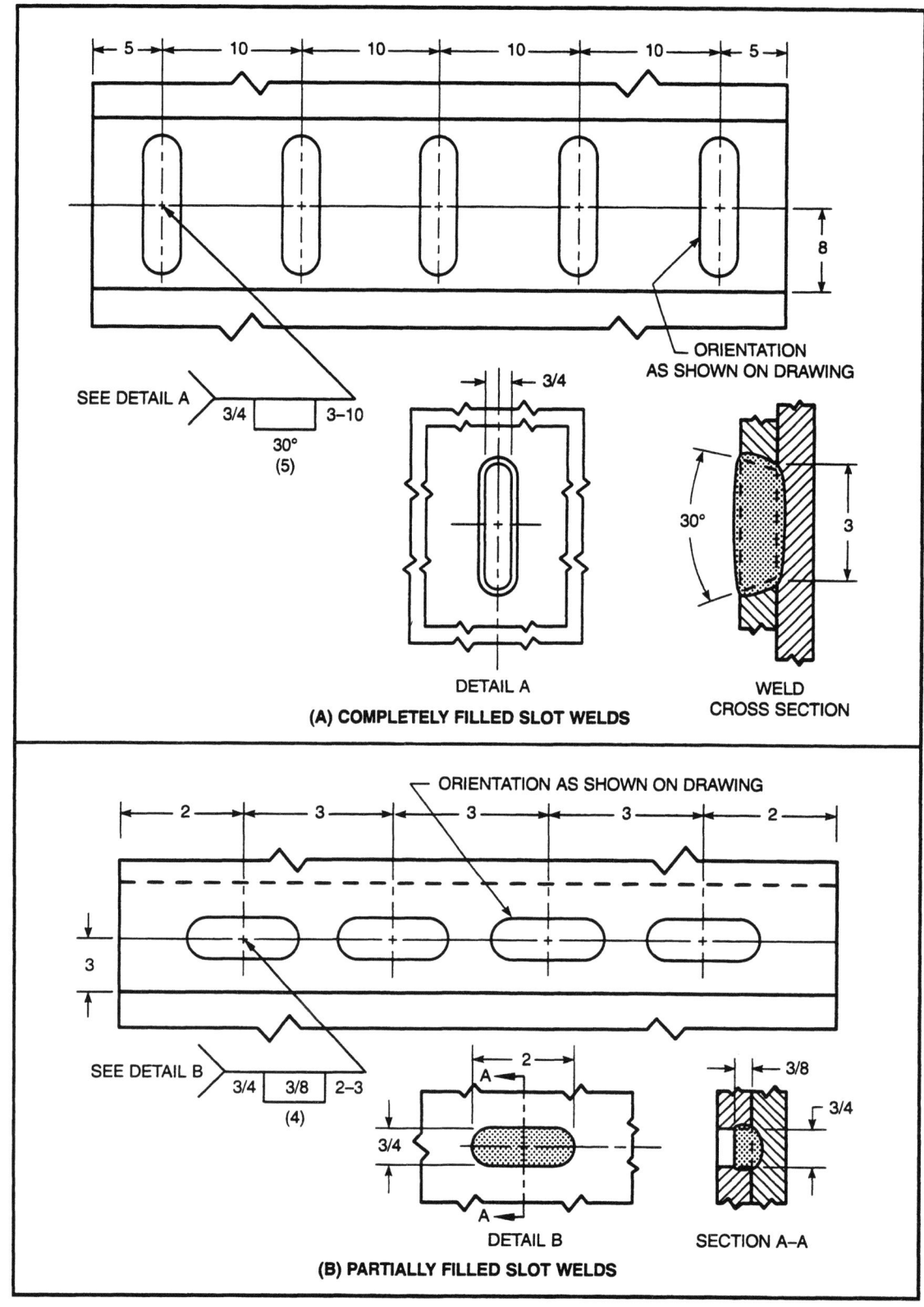

Figure 38—Applications of Information to Slot Weld Symbols

8. Spot Welds

8.1 General

8.1.1 Arrow-Side, Other-Side Significance. The spot weld symbol, relative to its location on the reference line, may or may not have arrow-side member or other-side member significance (see 3.1.2, 3.1.3, and Figure 39).

8.1.1.1 Arrow-Side Member. For those welding processes for which arrow-side member significance is applicable, the arrow-side member shall be indicated by placing the spot weld symbol below the reference line with the arrow pointing to this member [see Figures 1 and 39(A)].

8.1.1.2 Other-Side Member. For those welding processes for which other-side member significance is applicable, the other-side member shall be indicated by placing the spot weld symbol above the reference line [see Figure 39(B)].

8.1.1.3 No Side Significance. For those welding processes for which no arrow-side or other-side significance is applicable, the spot weld symbol shall be centered on the reference line [see 3.1.3 and Figure 39(C)].

8.1.2 Dimension Location. Dimensions shall be specified on the same side of the reference line as the spot weld symbol, or all dimensions shall be shown on either side when the spot weld symbol has no arrow-side or other-side significance (see Figures 39 and 40).

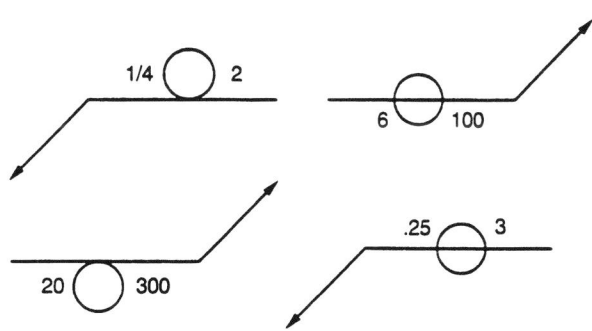

8.1.3 Welding Process Reference. The process reference shall be indicated in the tail of the welding symbol (see 3.11.1 and Figures 40 and 41).

8.1.4 Projection Welds. The projection weld symbol shall be used with the projection welding process reference in the tail of the welding symbol. The projection weld symbol shall be placed above or below (not center on) the reference line to designate which member receives the embossment in accordance with the location conventions given in 3.1.2 (see Figure 41).

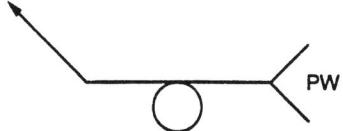

8.2 Size or Strength of Spot Welds. Spot welds shall be specified by either size or strength to the left of the spot weld symbol as follows:

8.2.1 Size. The size of a spot weld shall be specified, in inches or millimeters, as the diameter of the weld at the faying surfaces of the members [see Figure 40(A)].

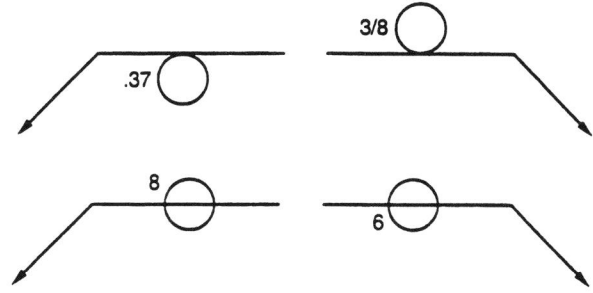

8.2.2 Strength. The shear strength of a spot weld shall be specified in pounds or newtons [see Figure 40(B)].

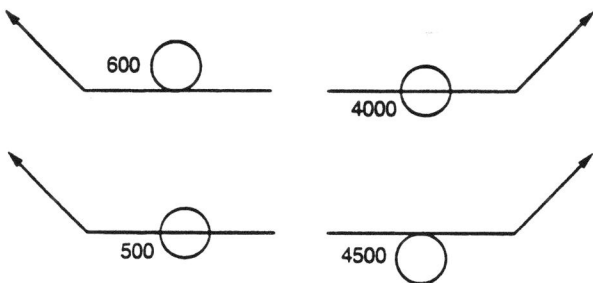

8.3 Spacing of Spot Welds. The pitch (center-to-center distance) of spot welds in a straight line shall be specified to the right of the weld symbol [see Figure 40(C)].

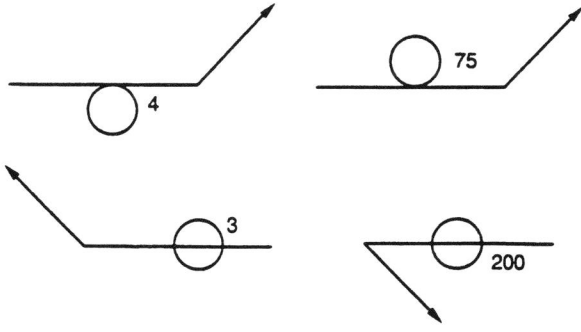

8.4 Number of Spot Welds

8.4.1 Number Specified. When a definite number of spot welds is desired in a joint, the number shall be specified in parentheses on the same side of the reference line as the spot weld symbol. The number may be either above or below the weld symbol when there is no other-side member significance and the symbol is centered on the reference line [see Figure 40(C), (D), (E), and (F)].

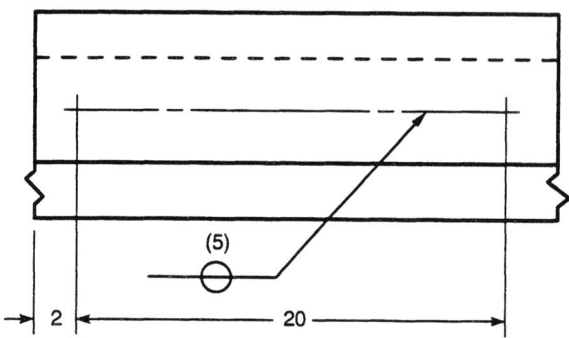

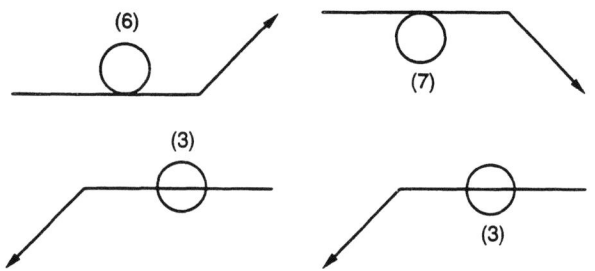

8.6 Contours and Finishing of Spot Welds

8.6.1 Contours Obtained by Welding. When the exposed surface of either member in a spot welded joint is to be welded with approximately a flush or convex face without postweld finishing, that surface shall be specified by adding the flush or convex contour symbol to the welding symbol (see 3.12).

8.4.2 Grouped Spot Welds. A group of spot welds may be located on a drawing by intersecting centerlines. The arrow shall point to at least one of the centerlines passing through each weld location. When spot welds are to be randomly located in a group, the area in which they are to be applied shall be clearly indicated [see Figure 40(E)].

8.6.2 Contours Obtained by Postweld Finishing. Spot welds whose faces are to be finished approximately flush or convex by postweld finishing, shall be specified by adding both the appropriate contour and finishing symbols to the welding symbol (see 3.13). Welds that require a flat but not flush surface require an explanatory note in the tail of the welding symbol.

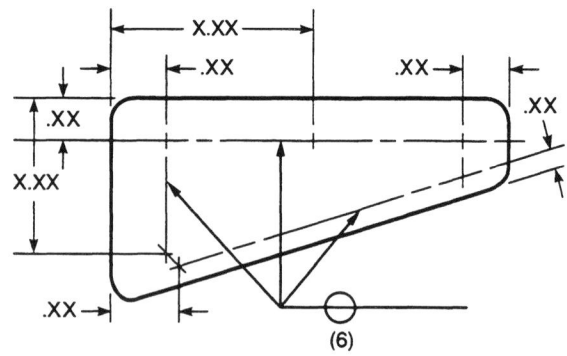

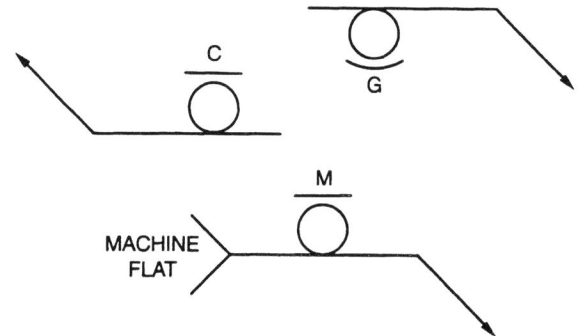

8.5 Extent of Spot Welding. When spot welds extend less than the distance between abrupt changes in the direction of welding, or less than the full length of the joint (see 3.9), the desired extent shall be dimensioned on the drawing [see Figure 40(D)].

8.7 Multiple-Member Spot Welds. When one or more members are included between the two outer members in a spot welded joint, the spot weld symbol for the two outer members shall be used (see Figure 42).

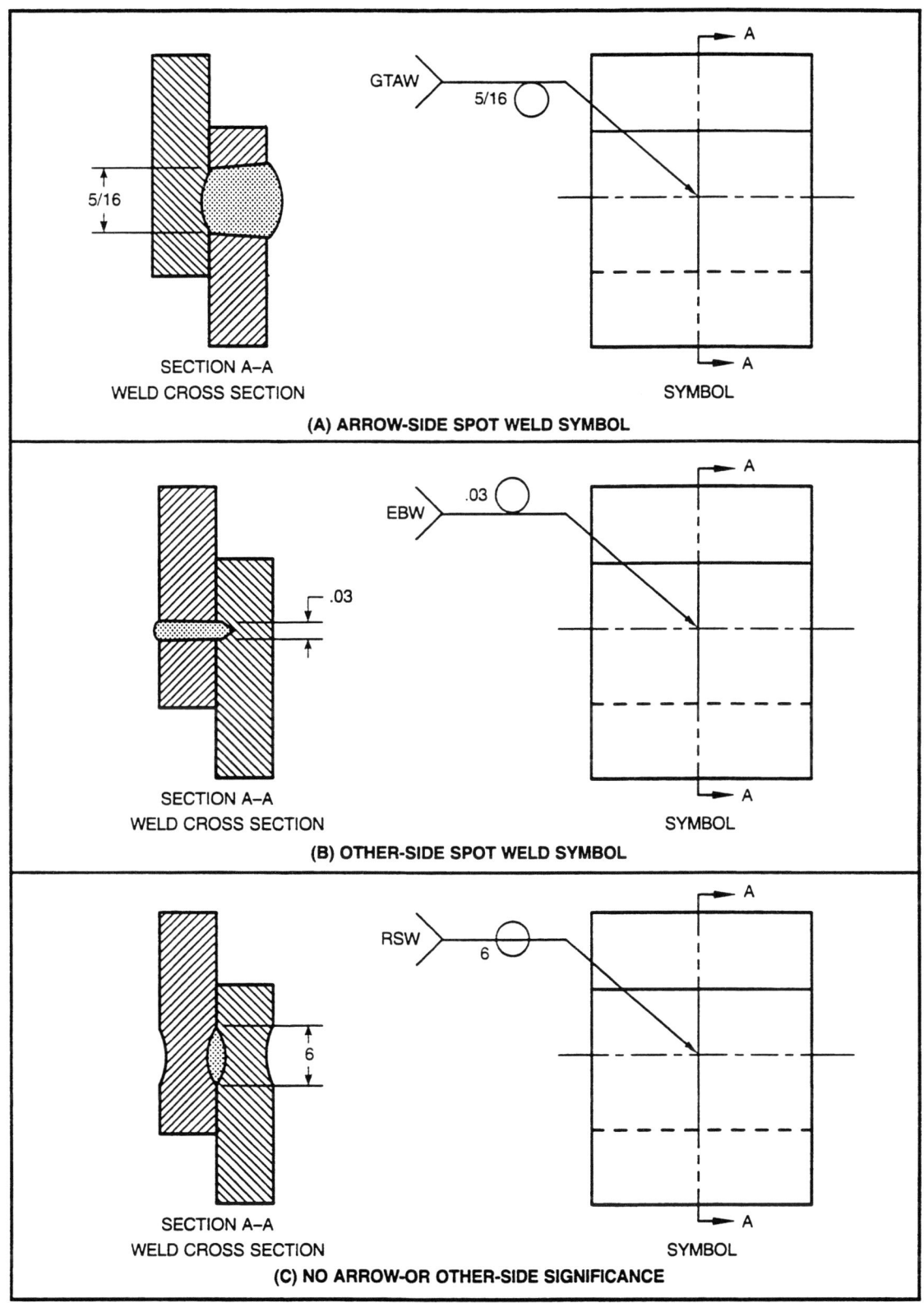

Figure 39—Applications of Spot Weld Symbol

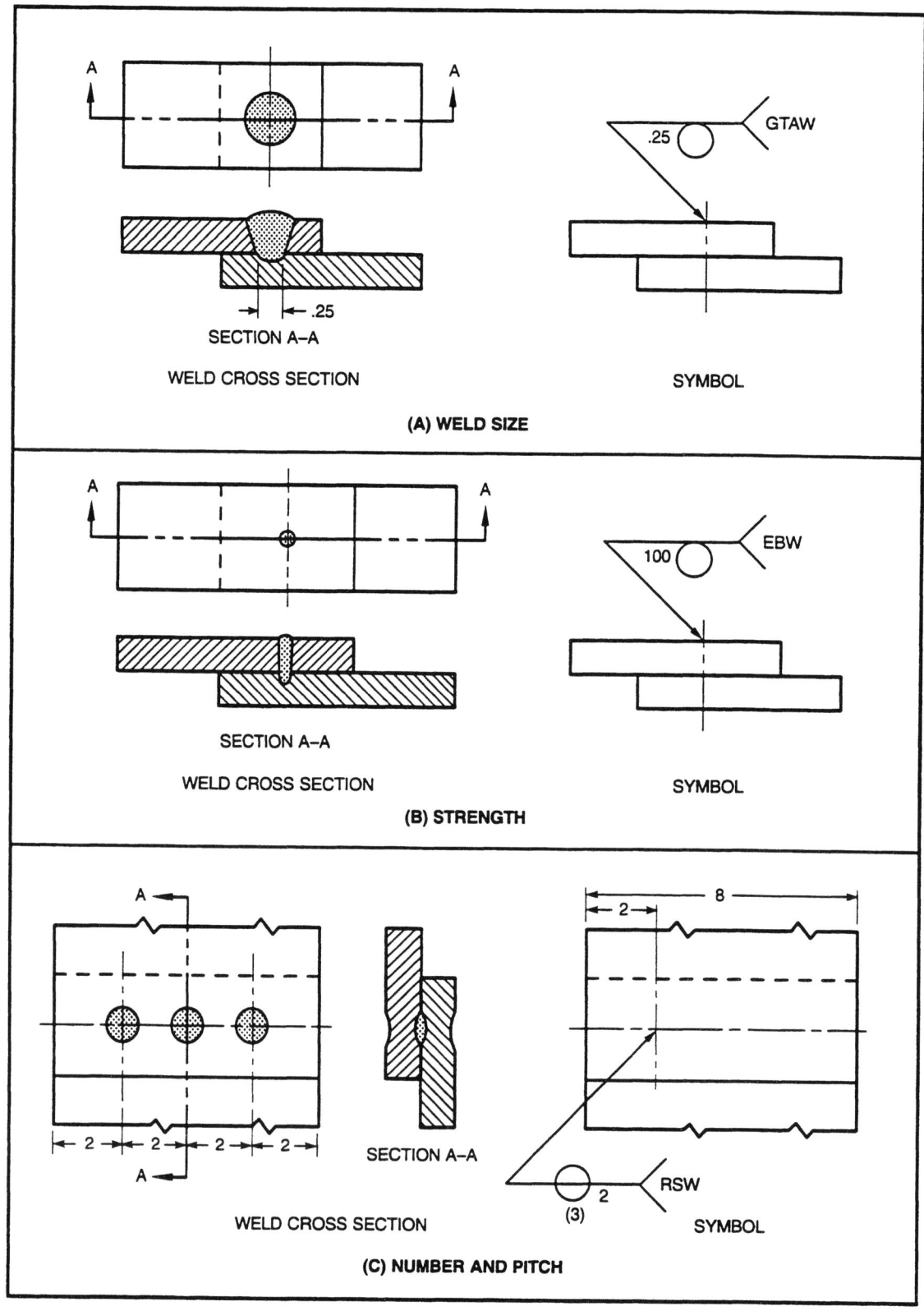

Figure 40—Applications of Information to Spot Weld Symbol

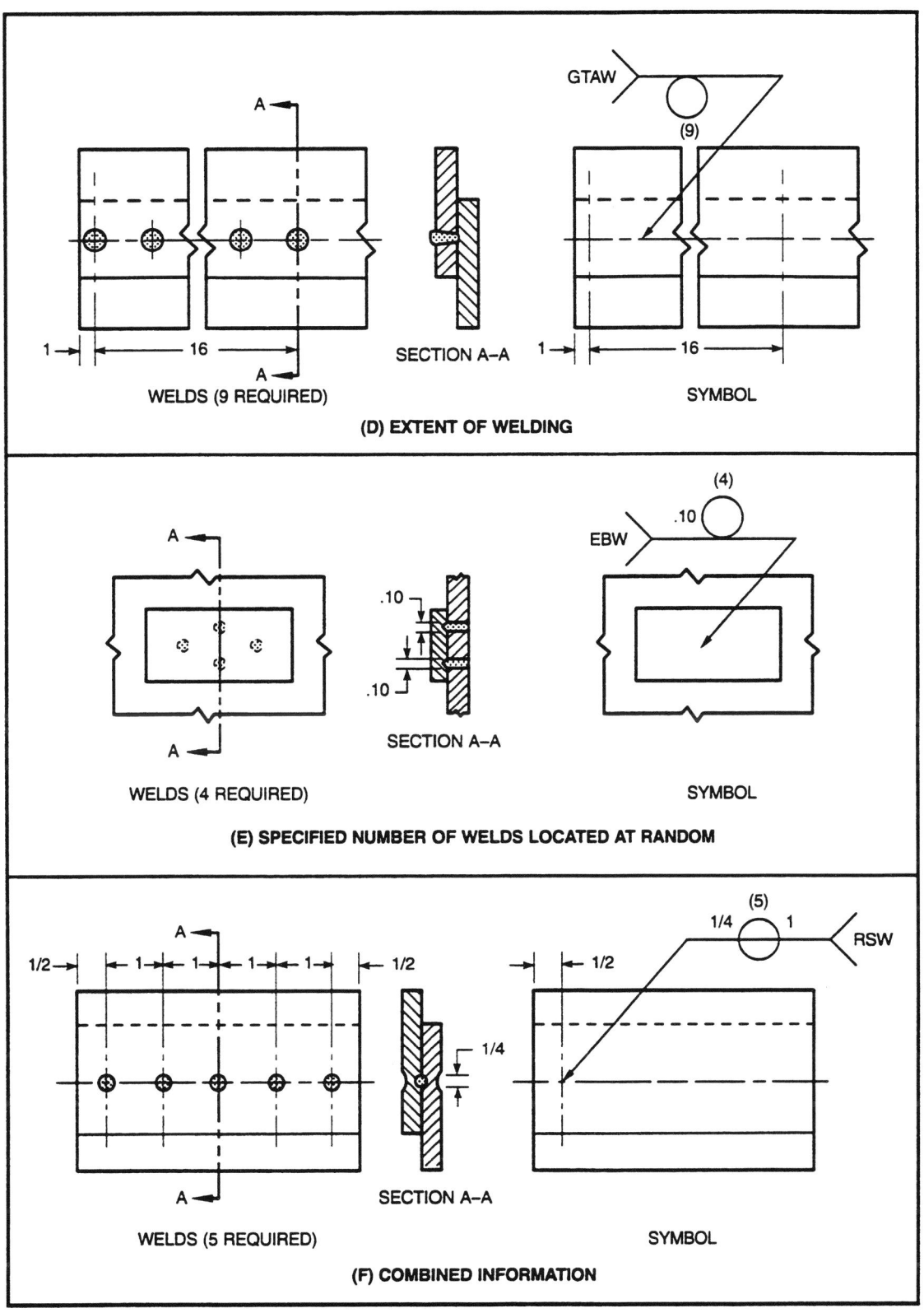

Figure 40 (Continued)—Applications of Information to Spot Weld Symbol

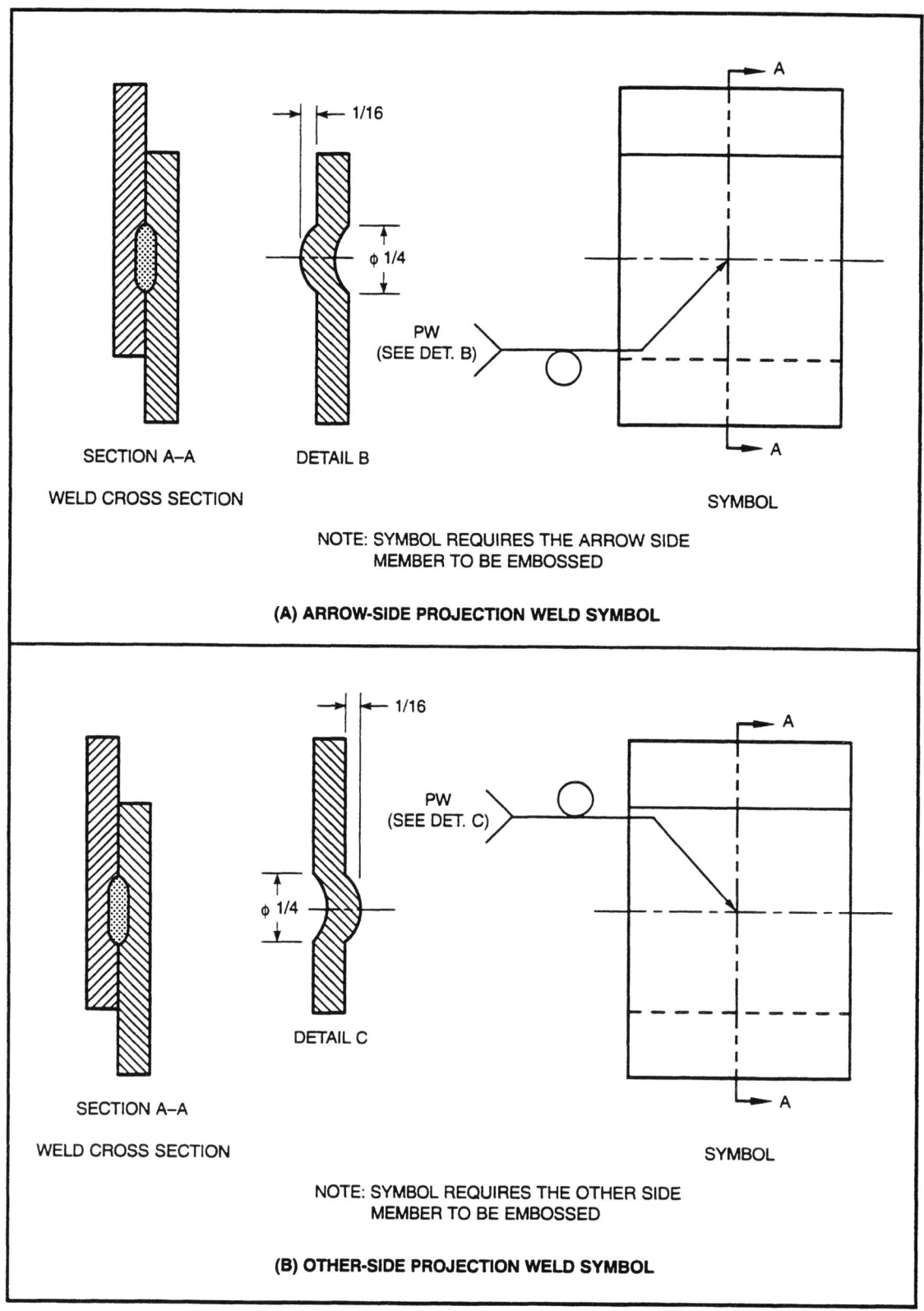

Figure 41—Applications of Projection Weld Symbol

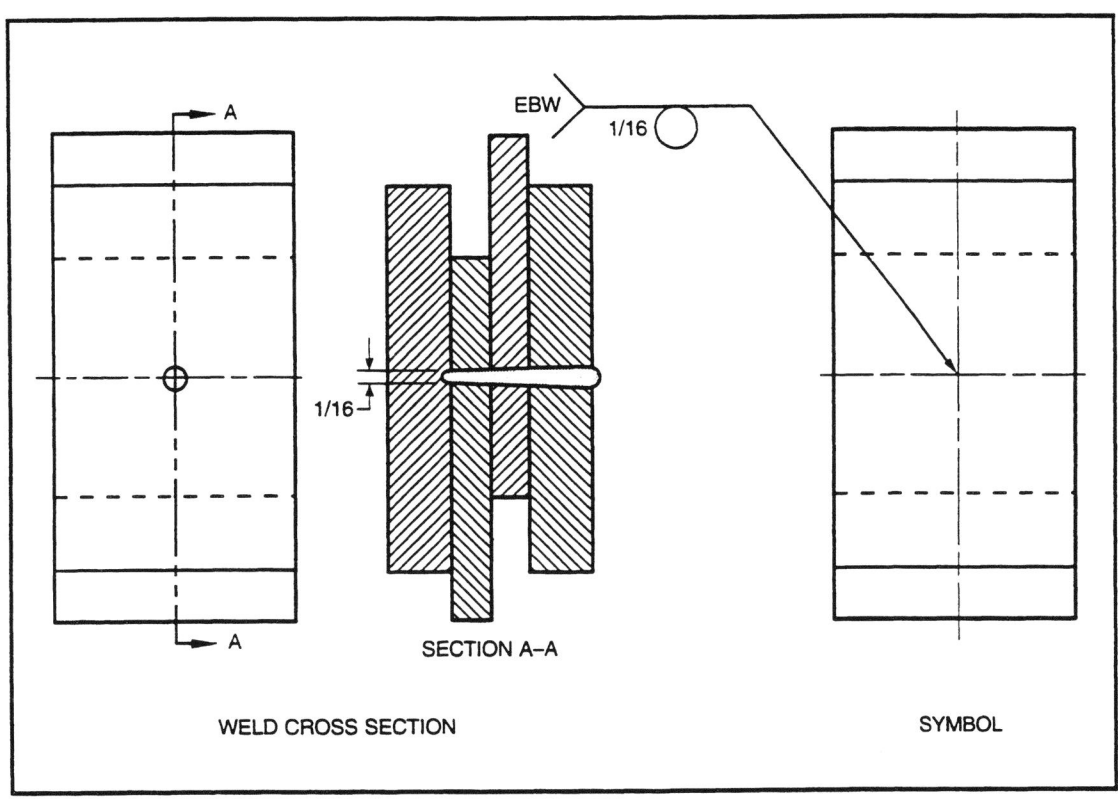

Figure 42—Multiple Member Spot Weld

9. Seam Welds

9.1 General

9.1.1 Arrow-Side, Other-Side Significance. The seam weld symbol, relative to its location on the reference line, may or may not have arrow-side member or other-side member significance (see 3.1.2, 3.1.3 and Figure 43).

9.1.1.1 Arrow-Side Member. For those welding processes for which arrow-side member significance is applicable, the arrow-side member shall be indicated by placing the seam weld symbol below the reference line with the arrow pointing to this member [see Figures 1 and 43(A)].

9.1.1.2 Other-Side Member. For those welding processes for which other-side significance is applicable, the other-side member shall be indicated by placing the seam weld symbol above the reference line [see Figure 43(B)].

9.1.1.3 No Side Significance. For those welding processes for which no arrow-side or other-side significance is applicable, the seam weld symbol shall be centered on the reference line [see 3.1.3 and Figure 43(C)].

9.1.2 Dimension Location. Dimensions shall be shown on the same side of the reference line as the weld symbol, or all dimensions shall be shown on either side when the seam weld symbol has no arrow-side or other-side significance (see Figure 44).

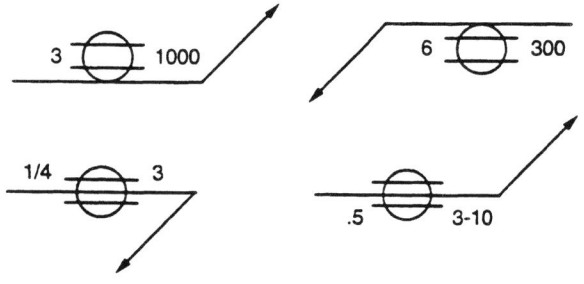

9.1.3 Welding Process Reference. The process reference shall be indicated in the tail of the welding symbol (see 3.11.1 and Figures 43–45).

9.2 Size and Strength of Seam Welds.
Seam welds shall be specified by either size or strength to the left of the seam weld symbol as follows:

9.2.1 Size. The size of a seam weld shall be specified, in inches or millimeters, as the width of the weld at the faying surfaces of the members [see Figure 44(A)].

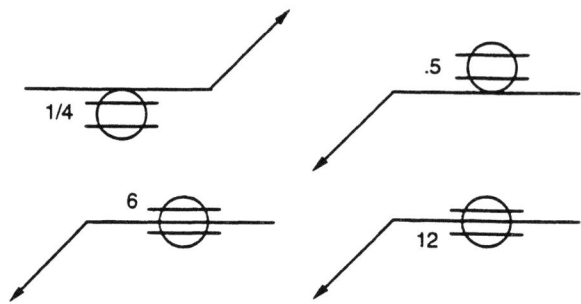

9.2.2 Strength. The shear strength of a seam weld shall be specified in pounds per linear inch or in newtons per millimeter [see Figure 44(B)].

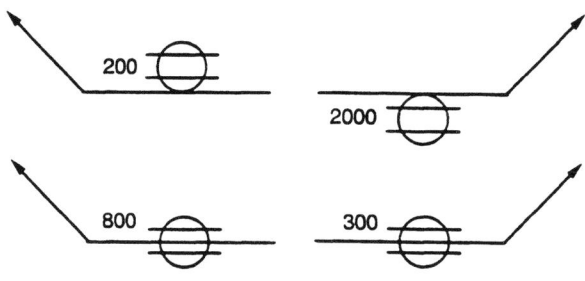

9.3 Length of Seam Welds

9.3.1 Dimension Location. The length of a seam weld shall be specified to the right of the weld symbol [see Figure 44(A) and (D)].

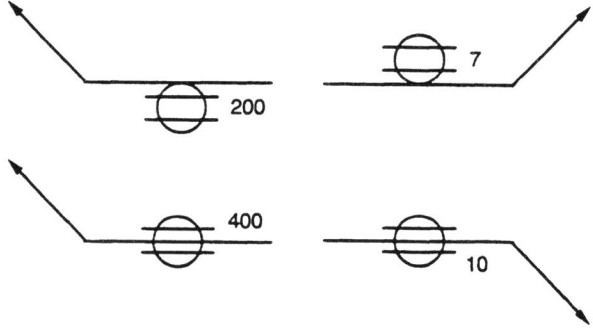

9.3.2 Abrupt Changes. When a seam weld extends the full distance between abrupt changes in the direction of welding (see 3.9), no length dimension need be specified on the welding symbol.

9.3.3 Specific Lengths. When a seam weld extends less than the distance between abrupt changes in the direction of welding, or less than the full length of the joint, the extent shall be dimensioned on the drawing [see 3.9 and Figure 44(C)].

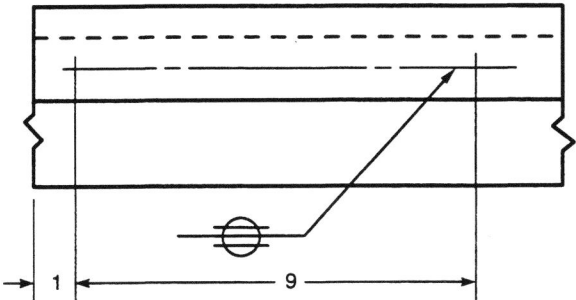

9.4 Dimensions of Intermittent Seam Welds

9.4.1 Pitch. The pitch of intermittent seam welds shall be specified as the distance between centers of the weld segments [see Figure 44(A) and (D)].

9.4.2 Pitch Dimension Location. The pitch of intermittent seam welds shall be specified to the right of the length dimension following a hyphen [see Figure 44(A) and (D)].

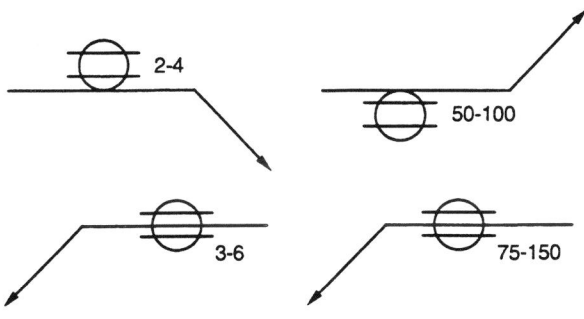

9.5 Number of Seam Welds. When a definite number of seam welds is desired in a joint, the number shall be specified in parentheses on the same side of the reference line as the weld symbol. The number shall be either above or below the weld symbol as appropriate [see Figure 44(D)].

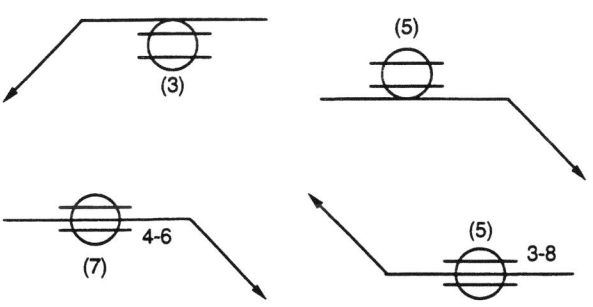

9.6 Orientation of Seam Welds

9.6.1 Intermittent Welds. Unless otherwise indicated, intermittent seam welds shall be interpreted as having length and pitch measured parallel to the weld axis [see Figure 44(A)].

9.6.2 Showing Orientation. When the orientation of seam welds is not as in 9.6.1, a detailed drawing shall be used to specify the weld orientation [see Figure 44(D)].

9.7 Contours and Finishing of Seam Welds

9.7.1 Contours Obtained by Welding. When the exposed surface of either member in a seam welded joint is to be welded with approximately a flush or convex face without postweld finishing, that surface shall be specified by adding the flush or convex contour symbol to the welding symbol (see 3.12).

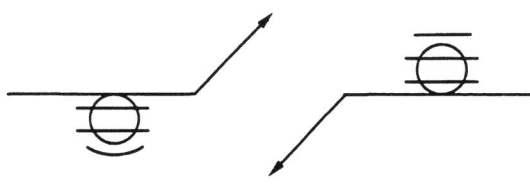

9.7.2 Contours Obtained by Postweld Finishing. Seam welds whose faces are to be finished approximately flush or convex by postweld finishing, shall be specified by adding both the appropriate contour and finishing symbols to the welding symbol (see 3.13). Welds that require a flat but not flush surface require an explanatory note in the tail of the welding symbol.

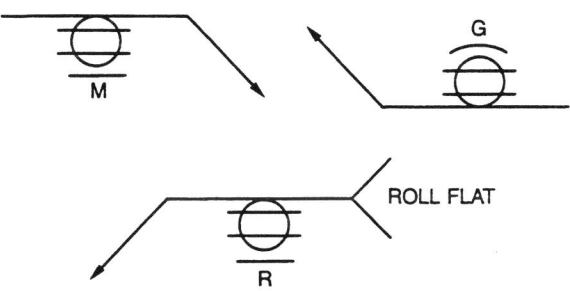

9.8 Multiple-Member Seam Welds. When one or more members are included between the two outer members in a seam welded joint, the seam weld symbol for the two outer members shall be used (see Figure 45).

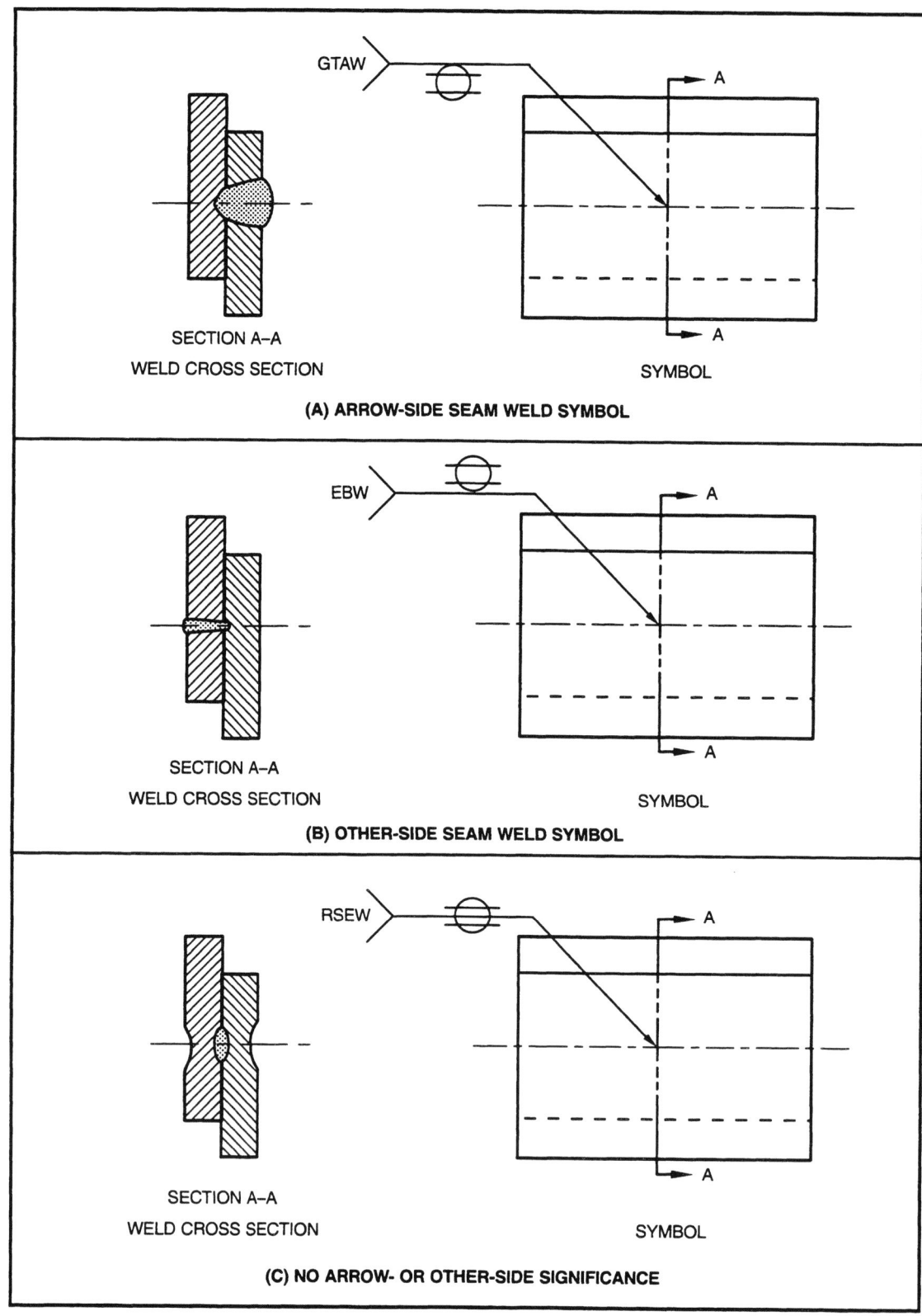

Figure 43—Applications of Seam Weld Symbol

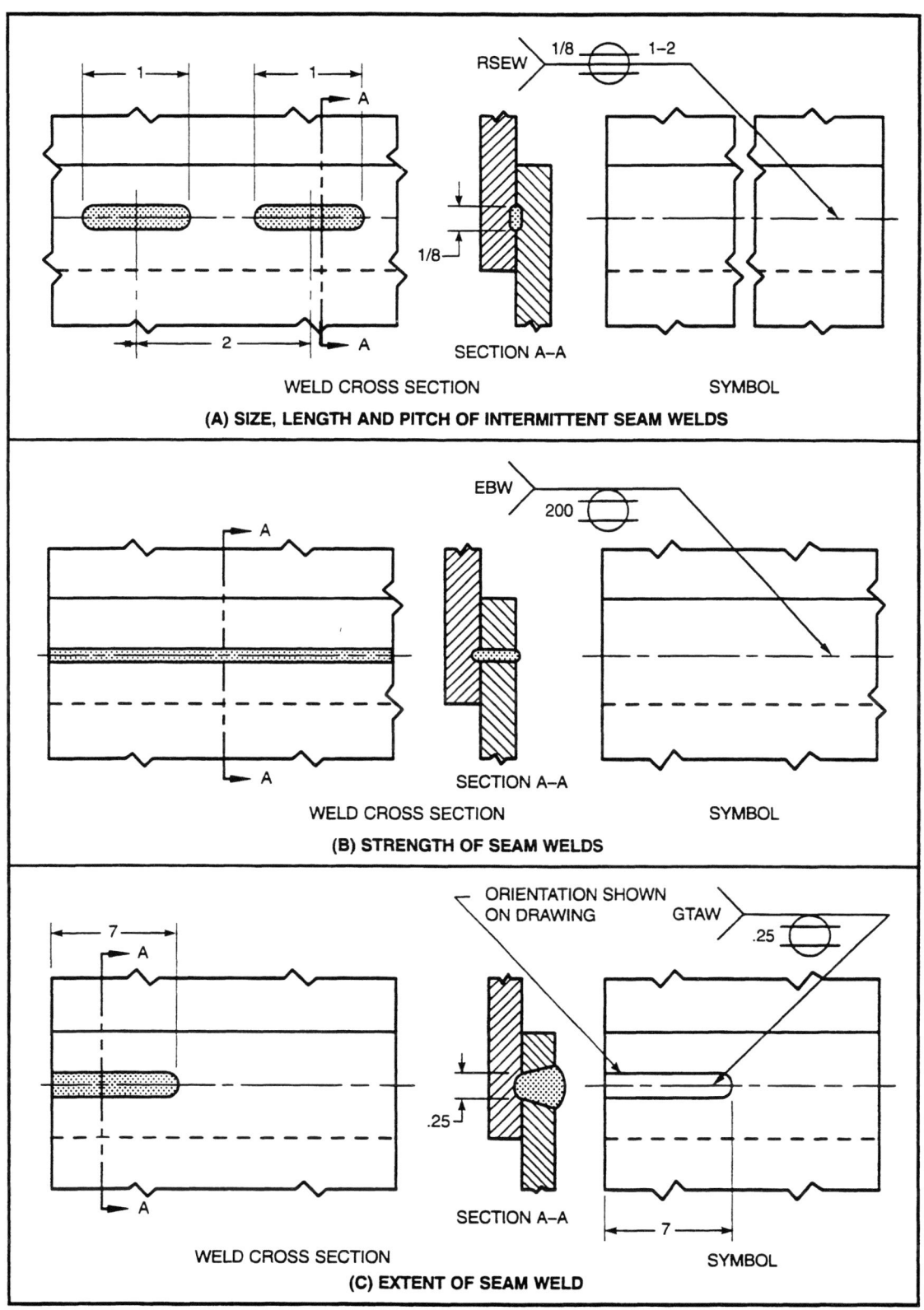

Figure 44—Applications of Information to Seam Weld Symbol

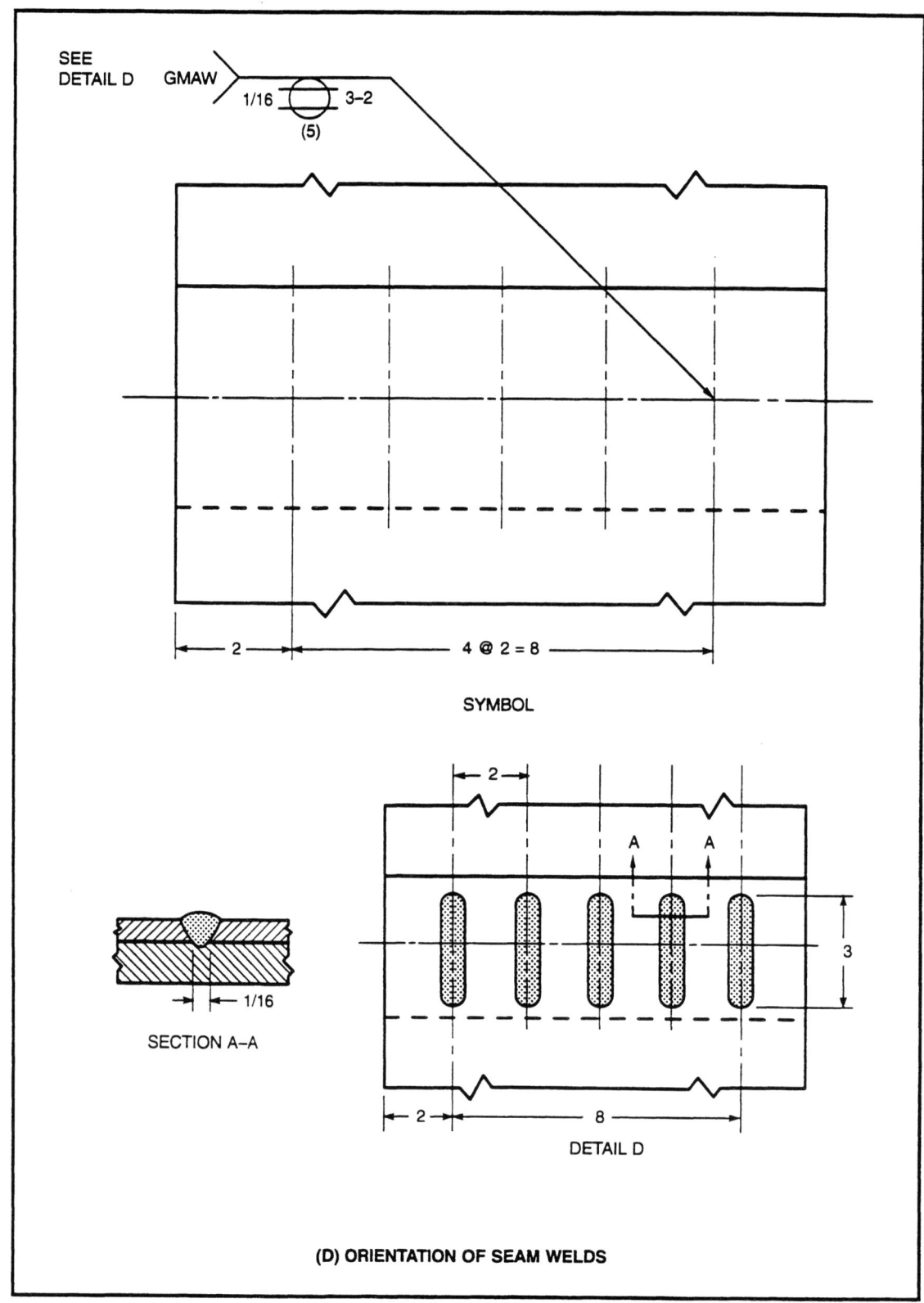

Figure 44 (Continued)—Applications of Information to Seam Weld Symbol

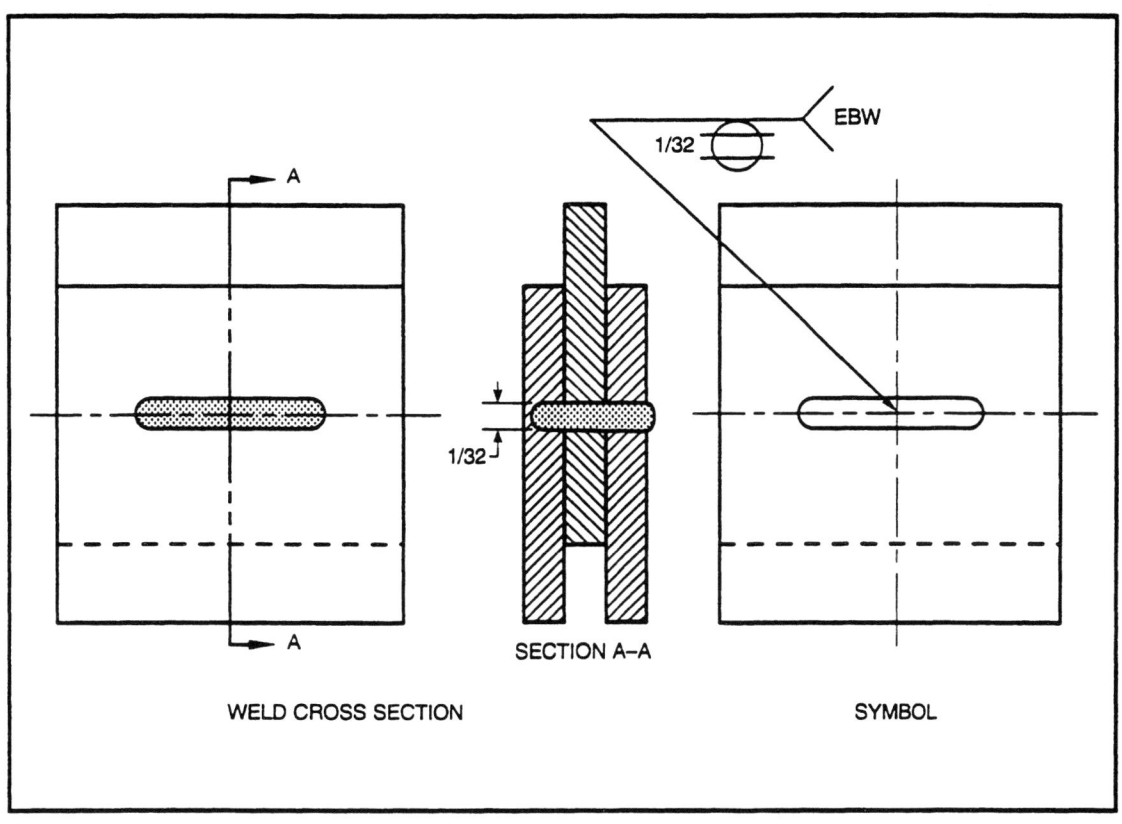

Figure 45—Multiple Member Seam Weld

10. Edge Welds

10.1 General. The edge weld symbol is used to specify edge welds on edge joints and flanged butt or flanged corner joints. The full thickness of the joint members must be fused. Flange dimensions are considered part of the drawing and not specified by the welding symbol. See Figure 46.

10.2 Edge Weld Size. When specified, the edge weld size shall be indicated by a dimension placed to the left of the edge weld symbol and on the same side of the reference line. If a specific edge weld size is not required, the dimension may be omitted [see Figure 46(A) and (B)].

10.3 Single- and Double-Edge Welds. Single-edge welds may be specified on edge, flanged butt, and flanged corner joints [see Figures 46(B), (C), and (D)]. Double-edge welds are only applicable to edge joints [see Figure 46(A)]. An edge weld may be combined with a flare-bevel or flare-V groove weld if welds are required on both sides of a flanged butt or flanged corner joint (see 4.2.10).

10.4 Edge Welds Requiring Complete Joint Penetration. Edge welds requiring complete joint penetration shall be specified for either flanged butt or flanged corner joints by the edge weld symbol with the melt-through symbol placed on the opposite side of the reference line [see Figures 46(E), (F), and (I)]. No size specification for the edge weld is necessary when combined with the melt-through symbol.

10.5 Edge Welds on Joints with More Than Two Members. Edge welds can be specified for edge joints, flanged butt joints, or flanged corner joints having more than two members by using the edge weld symbol in the same manner as for joints having two members [see Figures 46(G), (H), and (I).

10.6 Length of Edge Welds

10.6.1 Location. The length of an edge weld, when indicated on the welding symbol, shall be specified to the right of the weld symbol.

10.6.1.1 Full Length. When an edge weld is to extend for the full length of the joint, no length dimension need be specified on the welding symbol.

10.6.1.2 Specific Lengths. Specific lengths of edge welds and their location may be specified by symbols in conjunction with dimension lines.

10.6.1.3 Hatching. Hatching may be used to graphically depict edge welds.

10.6.2 Changes in the Direction of Welding. Symbols for edge welds involving changes in direction of welding shall be in accordance with 3.9.2.

10.7 Intermittent Edge Welds

10.7.1 Pitch. The pitch of intermittent edge welds shall be the distance between the centers of adjacent weld segments on one side of the joint.

10.7.2 Pitch Dimension Location. The pitch of intermittent edge welds shall be specified to the right of the length dimension following a hyphen.

10.7.3 Chain Intermittent Edge Welds. Dimensions of chain intermittent edge welds shall be specified on both sides of the reference line. The segments of chain intermittent edge welds shall be opposite one another across the joint.

10.7.4 Staggered Intermittent Edge Welds. Dimensions of staggered intermittent edge welds shall be specified on both sides of the reference line, and the edge weld symbols shall be offset on opposite sides of the reference line as shown below. The segments of staggered intermittent edge welds shall be symmetrically spaced on both sides of the joint.

10.7.5 Extent of Welding. In the case of intermittent edge welds, additional weld lengths which are intended at the ends of the joint shall be specified by separate welding symbols and dimensioned on the drawing. When no weld lengths are intended at the ends of the joint, the unwelded lengths should not exceed the clear distance between weld segments and be so dimensioned on the drawing.

10.7.6 Location of Intermittent Welds. When the location of intermittent welds is not obvious, such as on a circular weld joint, it will be necessary to provide specific segment locations by dimension lines (see 4.4.1.2 and 5.3.1.2) or by hatching (see 4.4.1.3 and 5.3.1.3).

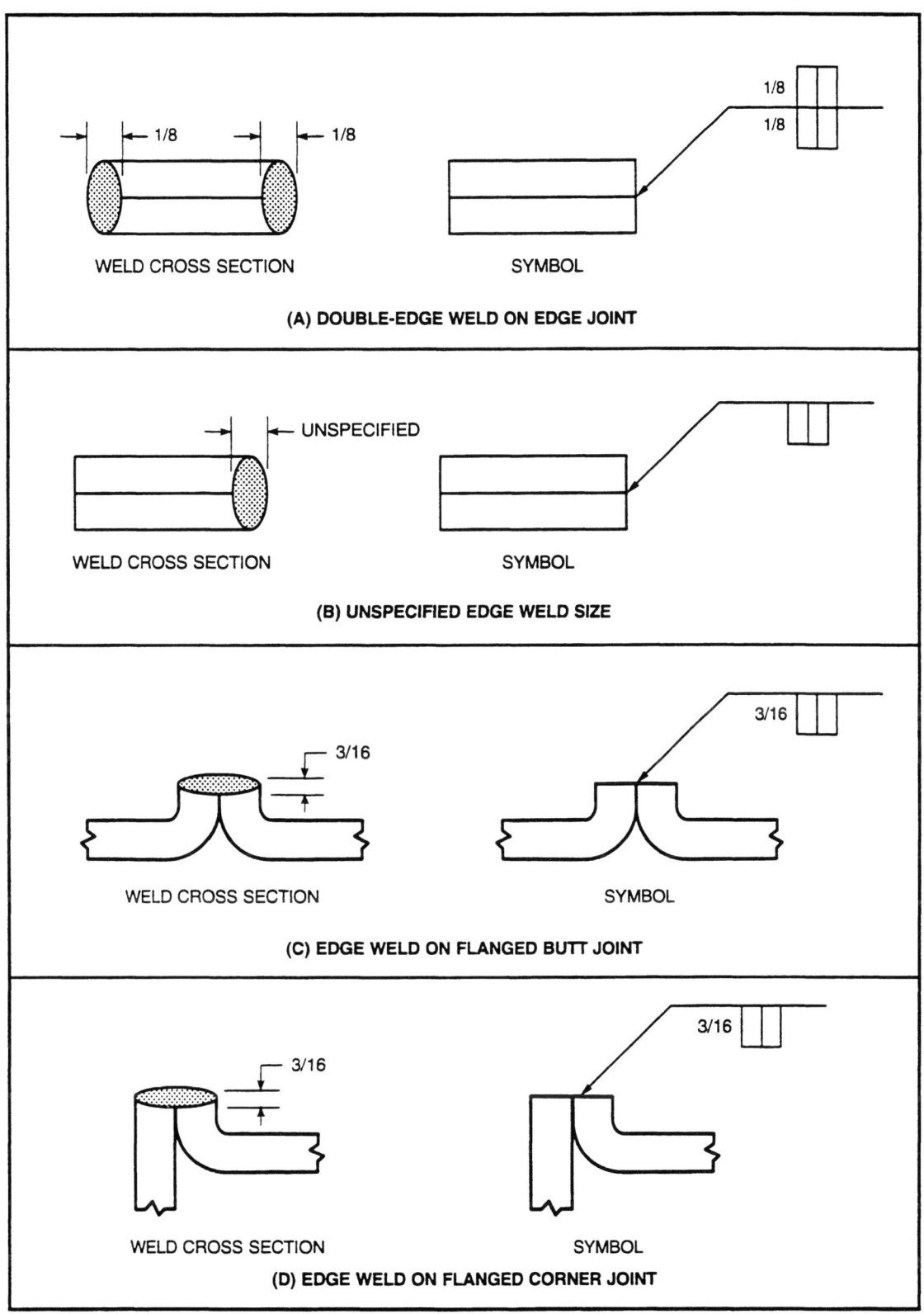

Figure 46—Applications of Edge Weld Symbols

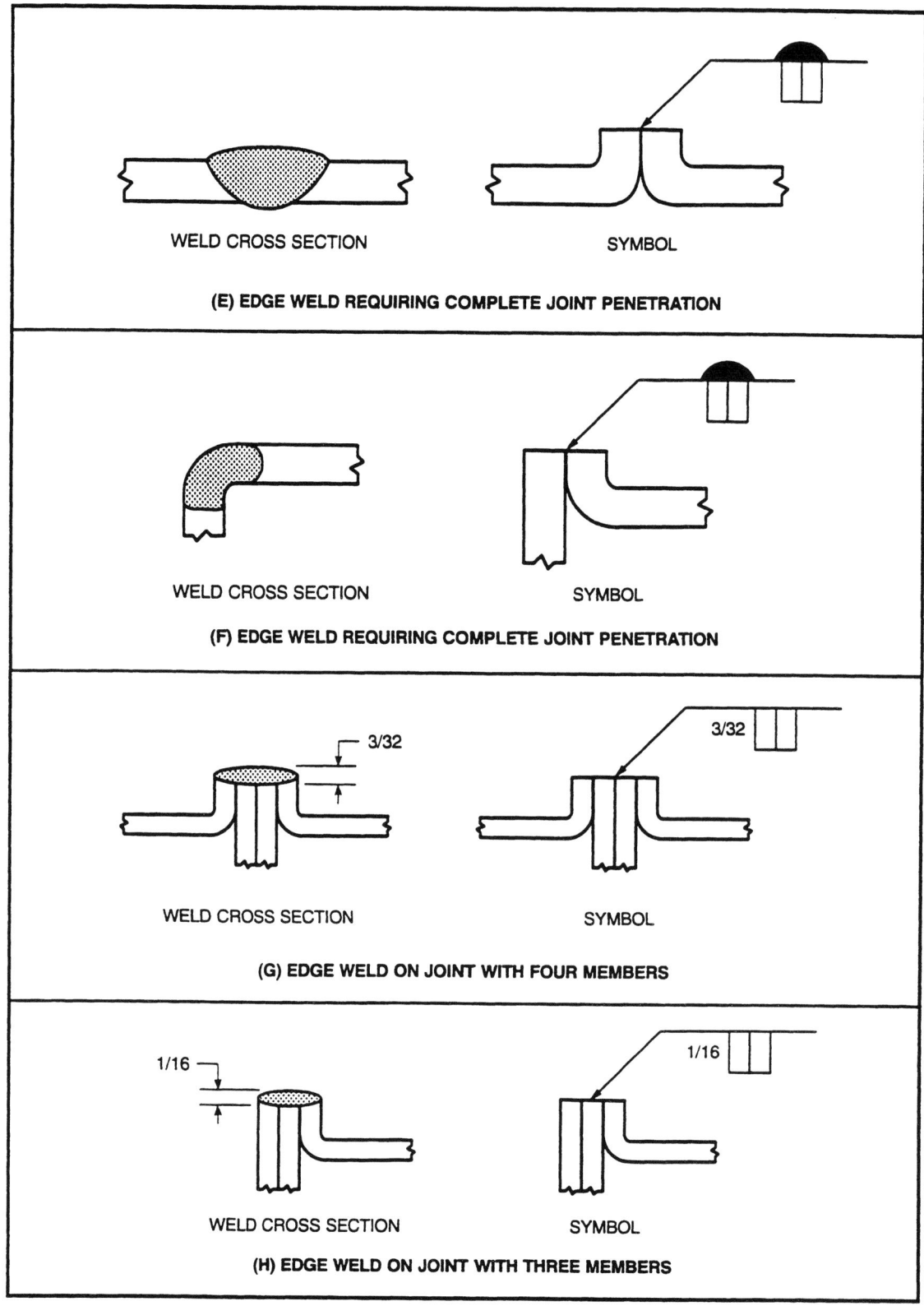

Figure 46 (Continued)—Applications of Edge Weld Symbols

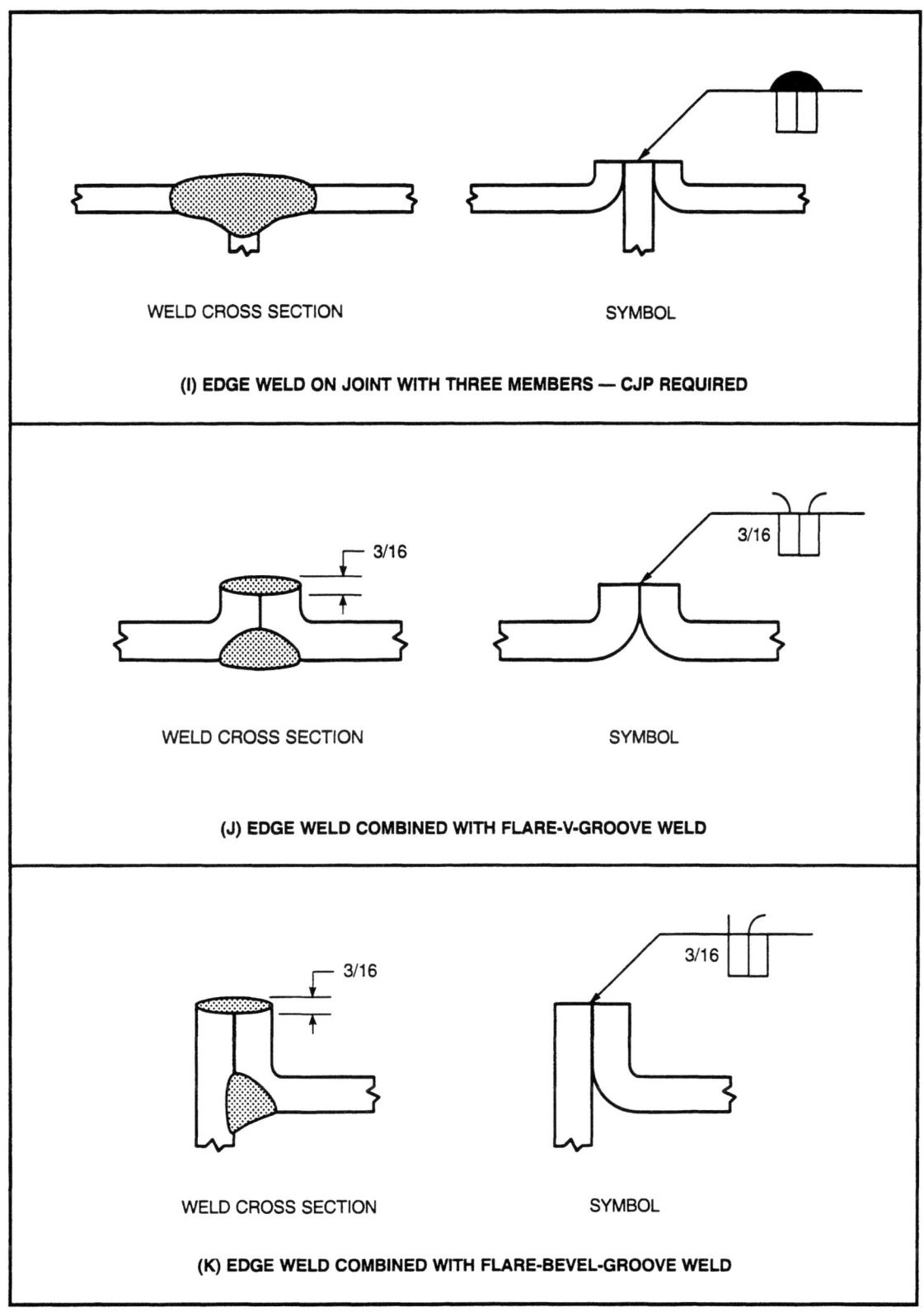

Figure 46 (Continued)—Applications of Edge Weld Symbols

11. Stud Welds

11.1 Side Significance. The stud weld symbol has arrow-side significance only. The symbol shall be placed below the reference line, and the arrow shall point clearly to the surface to which the stud is to be welded.

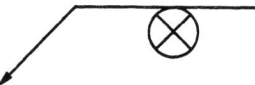

11.2 Stud Size. The required diameter of the stud shall be specified to the left of the weld symbol (see Figure 47).

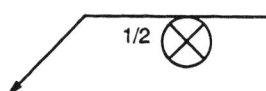

11.3 Spacing of Stud Welds. The pitch (center-to-center distance) of stud welds in a straight line shall be specified to the right of the weld symbol (see Figure 47). The spacing of stud welds in any configuration other than a straight line shall be dimensioned on the drawing.

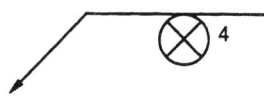

11.4 Number of Stud Welds. The number of stud welds shall be specified in parentheses below the stud weld symbol (see Figure 47)

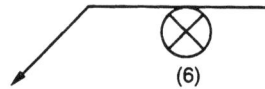

11.5 Dimension Location. Dimensions shall be placed on the same side of the reference line as the stud weld symbol (see Figure 47).

11.6 Location of First and Last Stud Welds. The location of the first and last stud weld in each single line shall be specified on the drawing (see Figure 47).

12. Surfacing Welds

12.1 Use of Surfacing Weld Symbol

12.1.1 Symbol Application. Surfacing, whether by single- or multiple-pass welds, shall be specified by the surfacing weld symbol (see Figure 48).

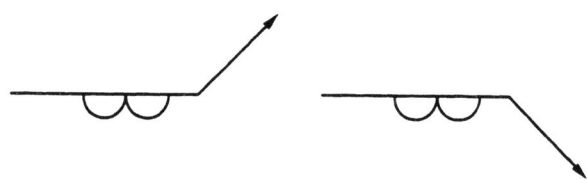

12.1.2 Arrow-Side Significance. The surfacing weld symbol does not indicate the welding of a joint and has arrow-side significance only. The symbol shall be placed below the reference line and the arrow shall point clearly to the surface on which the surfacing weld is to be deposited (see Figure 48).

12.1.3 Dimension Location. Dimensions used in conjunction with the surfacing weld symbol shall be placed on the same side of the reference line as the weld symbol [see Figure 48(A) and (C)].

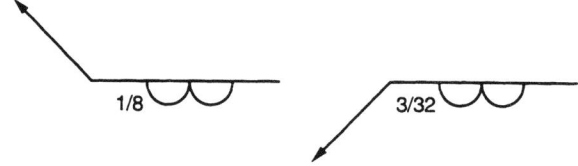

12.2 Size (Thickness) of Surfacing Welds

12.2.1 Minimum Thickness. The size (thickness) of a surfacing weld shall be specified by placing the dimension of the required thickness to the left of the weld symbol [see Figure 48(A) and (C)]. The direction of welding may be specified by a note in the tail of the welding symbol or indicated on the drawing.

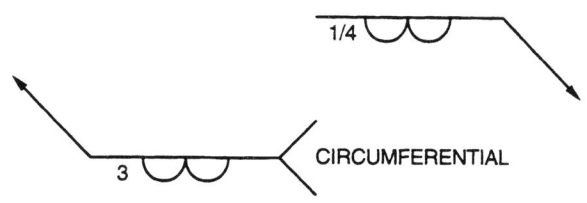

12.2.2 Multiple Layer. Multiple-layer surfacing welds may be specified by using multiple reference lines with the required size (thickness) of each layer placed to the left of the weld symbols. The direction of welding may be specified by an appropriate note in the tail of the welding symbol or indicated on the drawing [see Figure 48(C)].

12.3 Extent, Location, and Orientation of Surfacing Welds

12.3.1 Entire Area. No dimension other than size (thickness) is necessary to specify surfacing of the entire area of a plane or curved surface [see Figure 48(A)].

12.3.2 Portion of Area. When only a portion of a surface is to receive a surfacing weld, the extent, location, and orientation shall be shown on the drawing [see Figure 48(B) and (C)].

12.4 Surfacing a Previous Weld. Multiple reference lines may be used to specify a surfacing weld on the surface of a previously made weld (see 3.7).

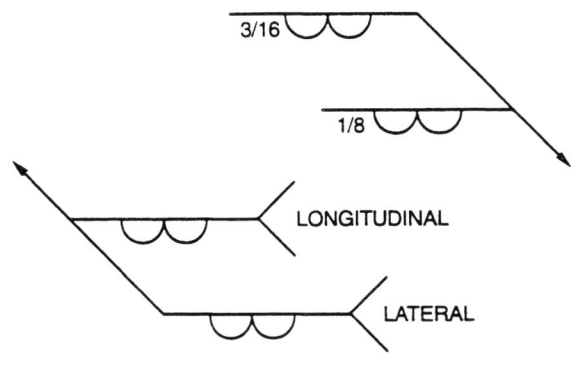

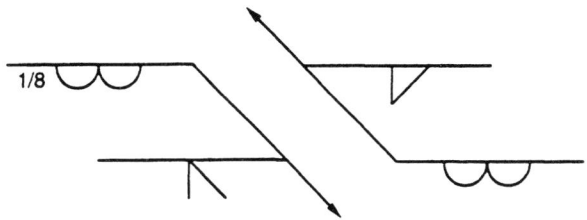

12.2.3 Unspecified Size (Thickness). When no specific thickness of a surfacing weld is required, the size dimension need not be included in the welding symbol [see Figure 48(B)].

12.5 Surfacing to Adjust Dimensions. The surfacing weld symbol may be used to specify a surfacing weld to correct assembly problems such as reducing excessive root openings [see Figure 48(D)].

Part B
Brazing Symbols

13. Brazed Joints

If no special preparation other than cleaning is required, only the arrow and reference line need be used with the brazing process indicated in the tail [see Figure 49(A)]. Applications of conventional welding symbols to brazed joints are illustrated in Figure 49(B) through (H). Figure 49(C), (D), (E), (G), and (H) show how joint clearances can be indicated. All symbols used for welding may also be used for brazing, where suitable.

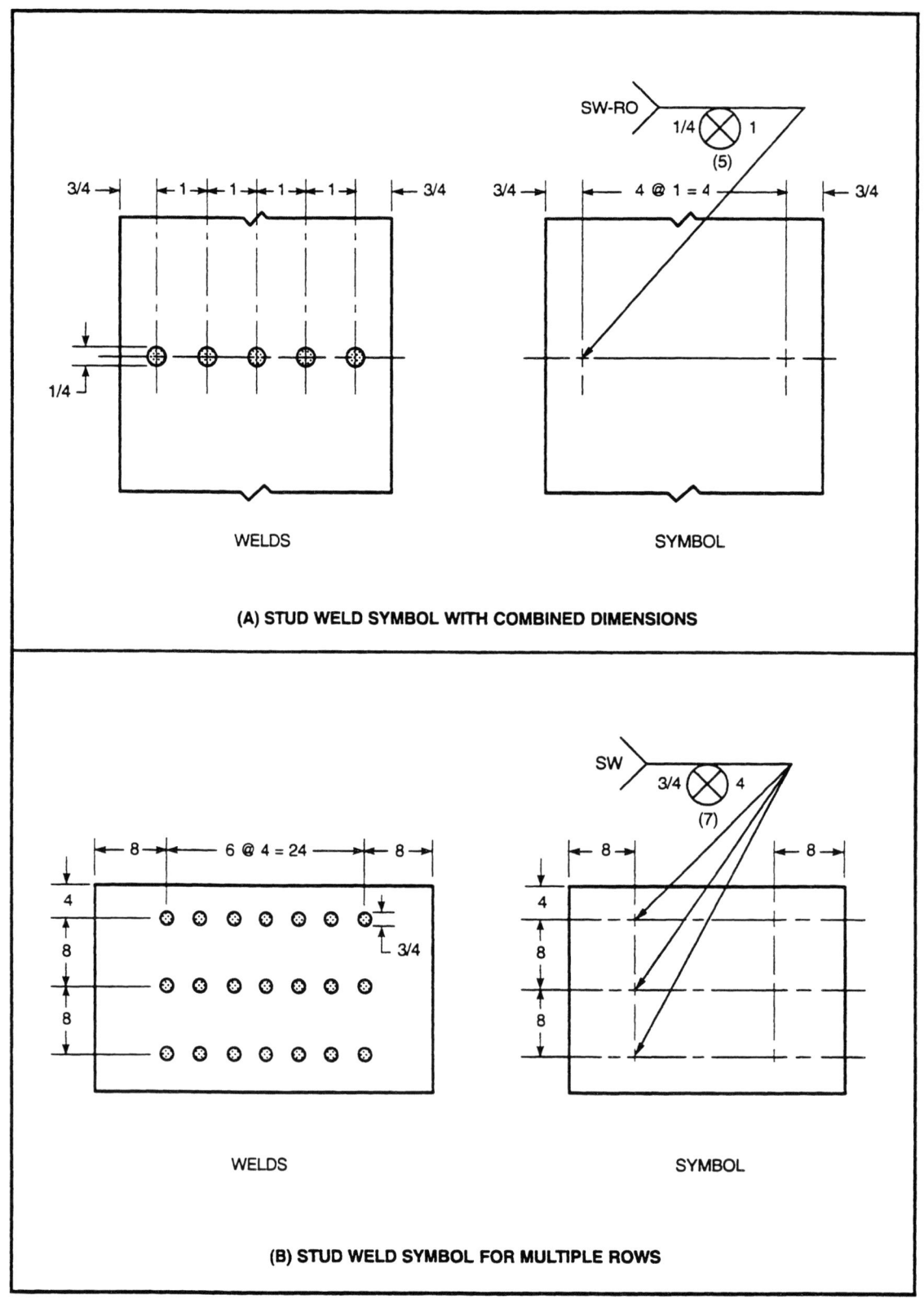

Figure 47—Applications of Stud Weld Symbol

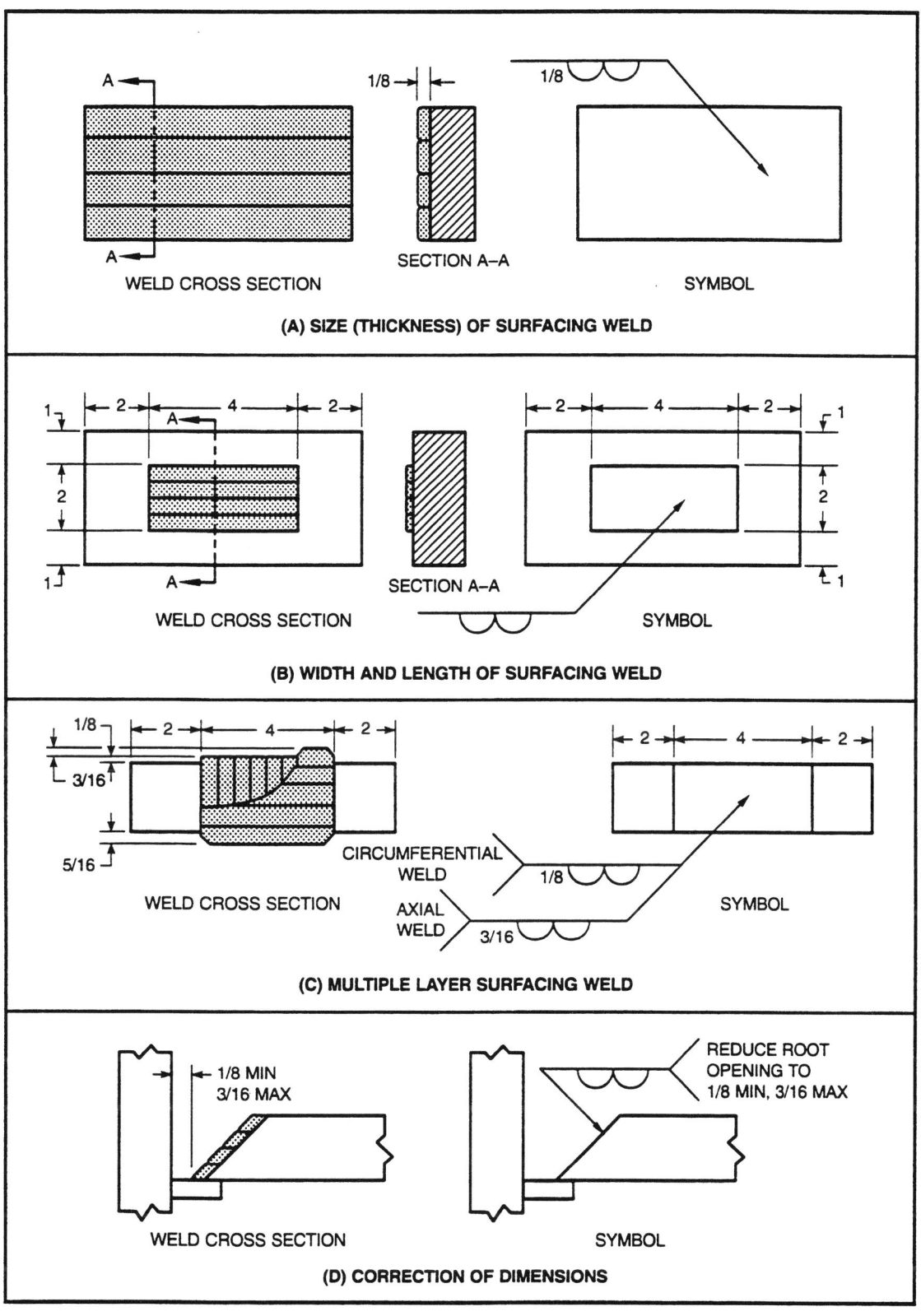

Figure 48—Applications of Surfacing Weld Symbol

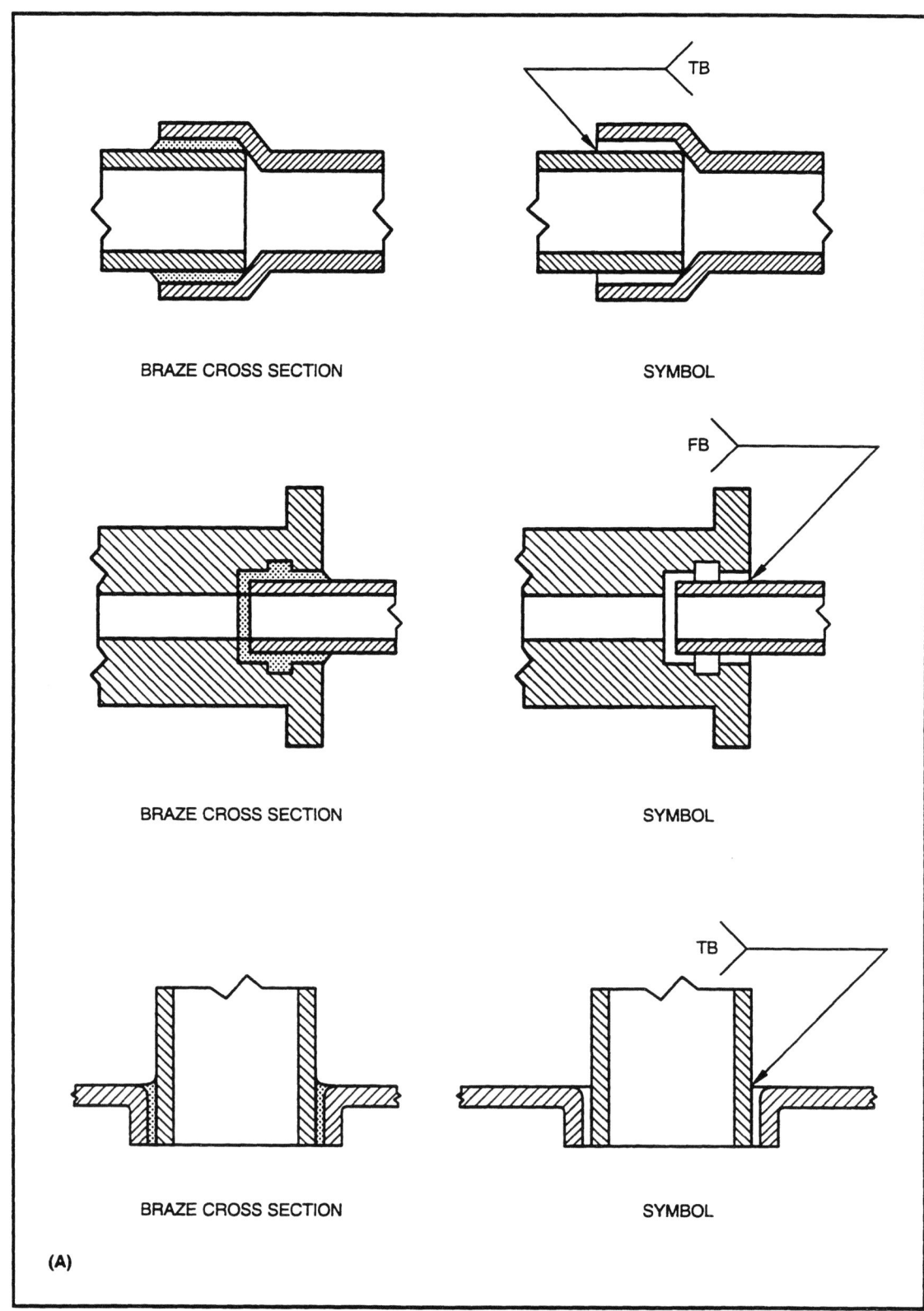

Figure 49—Applications of Brazing Symbols

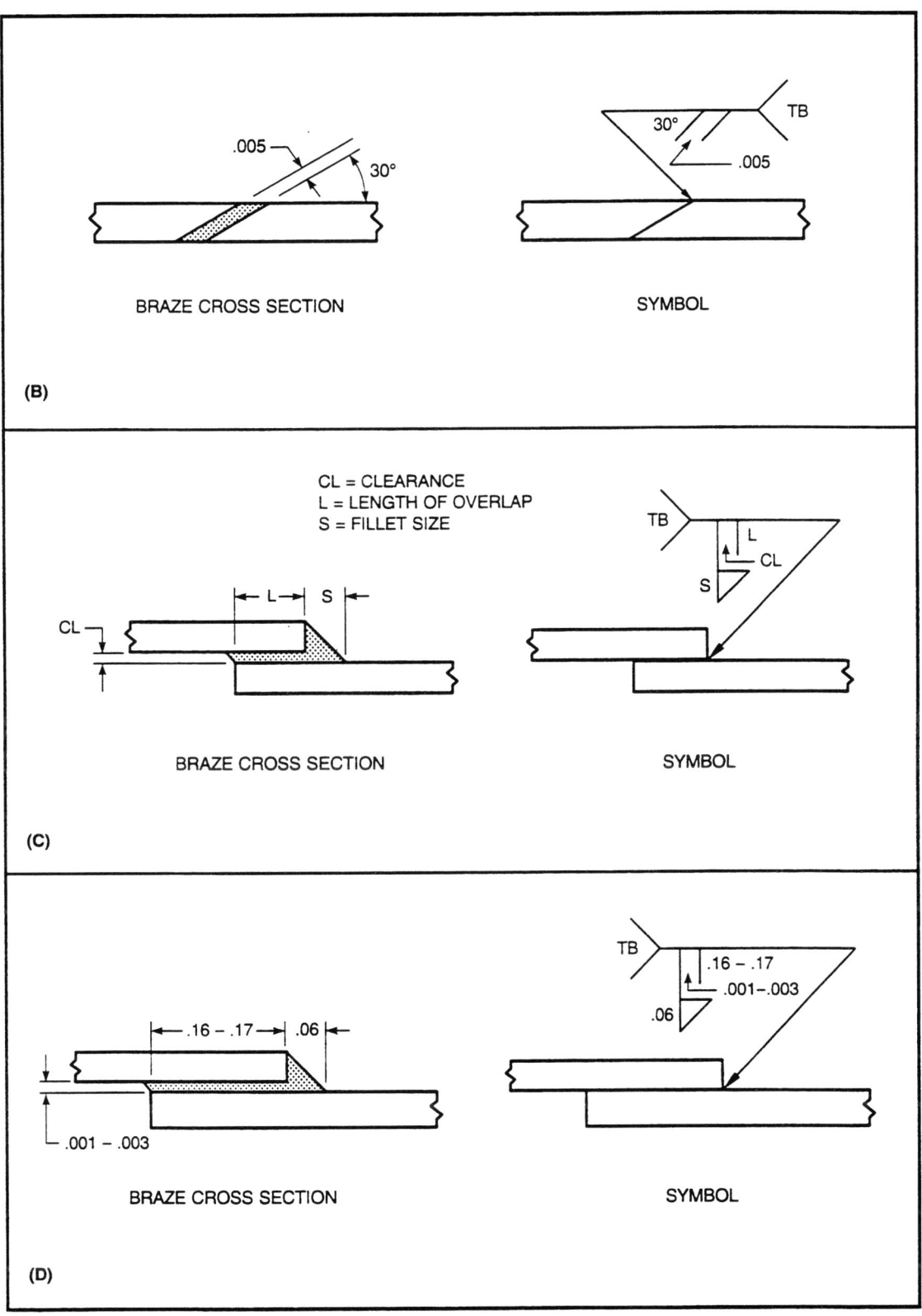

Figure 49 (Continued)—Applications of Brazing Symbols

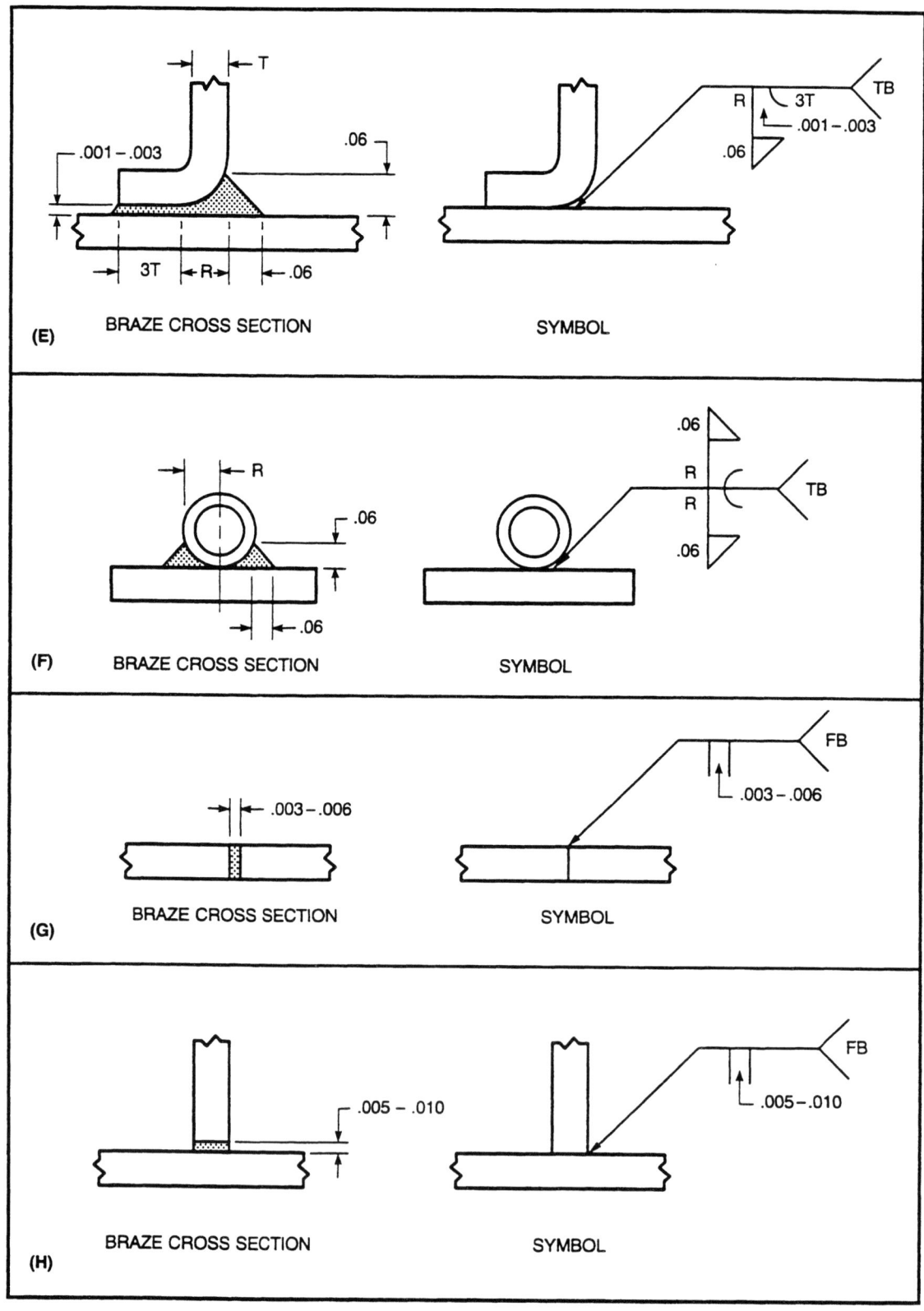

Figure 49 (Continued)—Applications of Brazing Symbols

Part C
Nondestructive Examination Symbols

14. Elements of the Nondestructive Examination Symbol

The examination symbol consists of the following elements:

(1) Reference line
(2) Arrow
(3) Examination method letter designations
(4) Extent and number of examinations
(5) Supplementary symbols
(6) Tail (specifications, codes, or other references)

14.1 Examination Method Letter Designations. Nondestructive examination methods shall be specified by use of the letter designation shown below.

Examination Method	Letter Designation
Acoustic emission	AET
Electromagnetic	ET
Leak	LT
Magnetic particle	MT
Neutron radiographic	NRT
Penetrant	PT
Proof	PRT
Radiographic	RT
Ultrasonic	UT
Visual	VT

14.2 Supplementary Symbols. Supplementary symbols to be used in nondestructive examination symbols shall be as follows:

EXAMINE ALL AROUND	FIELD EXAMINATION	RADIATION DIRECTION
⌀	⚑	✳

14.3 Standard Location of Elements of a Nondestructive Examination Symbol. The elements of a nondestructive examination symbol shall have standard locations with respect to each other as shown in Figure 50.

15. General Provisions

15.1 Location Significance of Arrow. The arrow shall connect the reference line to the part to be examined. The side of the part to which the arrow points shall be considered the arrow side of the part. The side opposite the arrow side of the part shall be considered the other side.

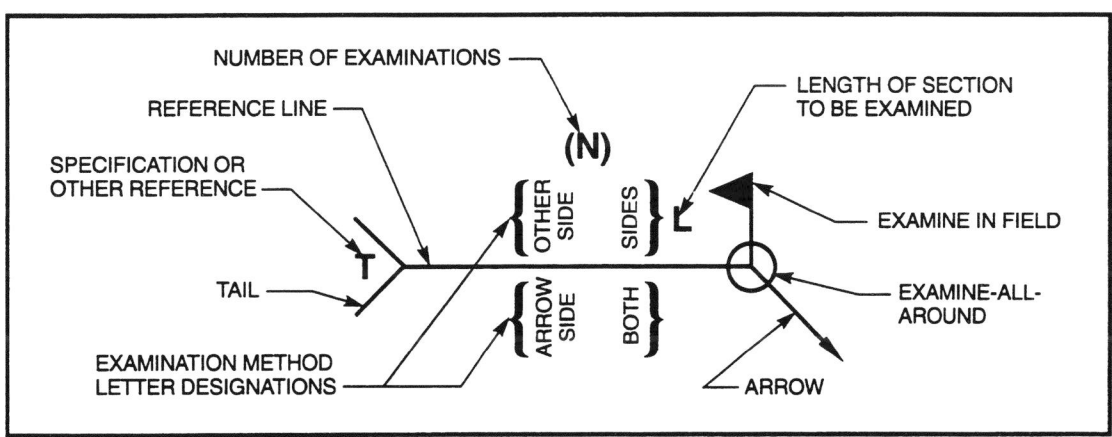

Figure 50—Standard Location of Elements

15.2 Location of Letter Designations

15.2.1 Location on Arrow Side. Examinations to be made on the arrow side of the part shall be specified by placing the letter designation for the selected examination method below the reference line.

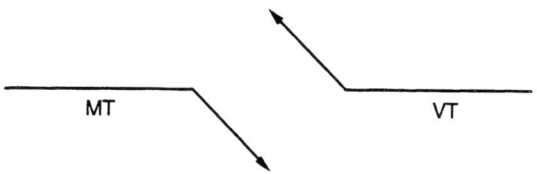

15.2.2 Location on the Other Side. Examinations to be made on the other side of the part shall be specified by placing the letter designation for the selected examination method above the reference line.

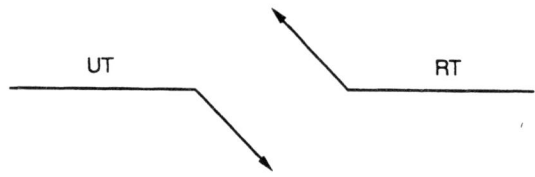

15.2.3 Location on Both Sides. Examinations to be made on both sides of the part shall be specified by placing the letter designation for the selected examination method on both sides of the reference line.

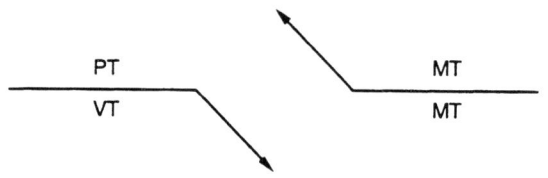

15.2.4 Location Centered on Reference Line. When the letter designation has no arrow- or other-side significance, or there is no preference from which side the examination is to be made, the letter designation shall be centered on the reference line.

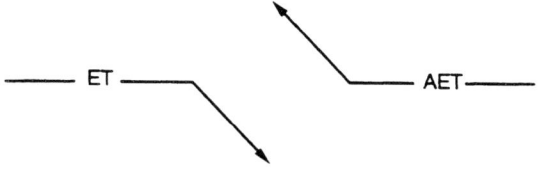

15.2.5 Examination Combinations. More than one examination method may be specified for the same part by placing the combined letter designations of the selected examination methods in the appropriate positions relative to the reference line. Letter designations for two or more examination methods, to be placed on the same side of the reference line or centered on the reference line, shall be separated by a plus sign.

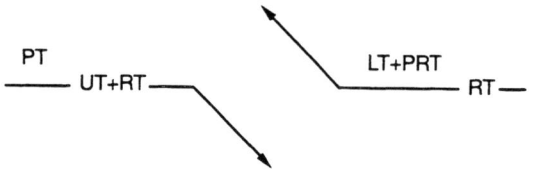

15.2.6 Welding and NDE Symbols. Nondestructive examination symbols and welding symbols may be combined.

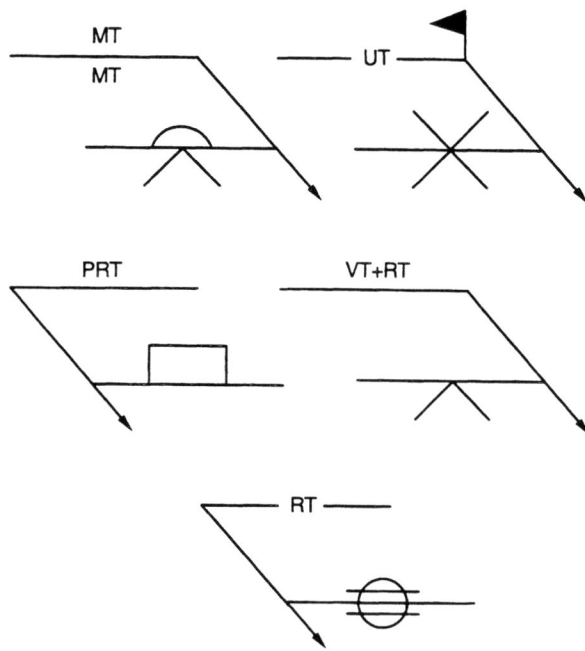

15.3 U.S. Customary and Metric Units. When it is required to specify dimensions with nondestructive examination symbols, the same system of units that is standard for the drawing shall be used. Dual dimensioning shall not be used on nondestructive examination symbols. If it is required to include conversions from metric to U.S. customary, or vice versa, a table of conversions may be included on the drawing. For guidance in drafting standards, reference is made to the ANSI Y14, *Drafting Manual*. For guidance on the use of metric (SI) units, reference is made to ANSI/AWS A1.1, *Metric Practice Guide for the Welding Industry*.

16. Supplementary Symbols

16.1 Examine-All-Around. Examinations required all around a weld, joint or part shall be specified by placing the examine-all-around symbol at the junction of the arrow and reference lines.

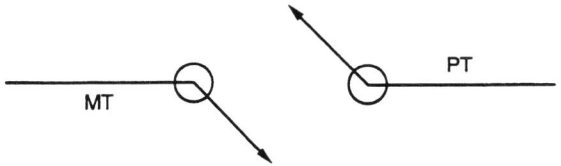

16.2 Field Examinations. Examinations required to be conducted in the field (not in a shop or at the place of initial construction) shall be specified by placing the field examination symbol at the junction of the arrow and reference lines.

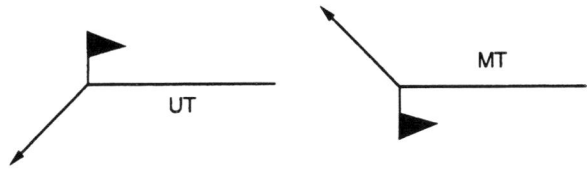

16.3 Radiation Direction. The direction of penetrating radiation may be specified by use of the radiation direction symbol drawn at the required angle on the drawing and the angle indicated, in degrees, to ensure no misunderstanding.

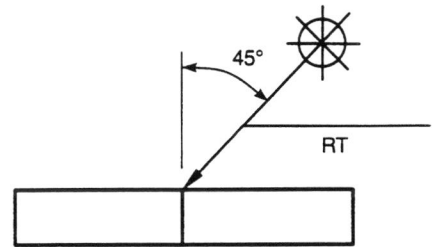

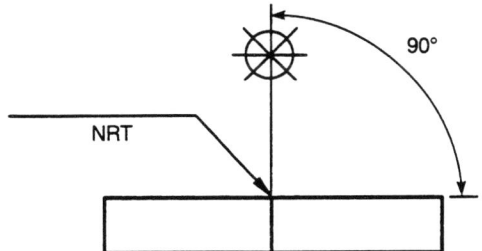

17. Specifications, Codes, and References

Information, applicable to the examination specified and which is not otherwise provided, may be placed in the tail of the nondestructive examination symbol.

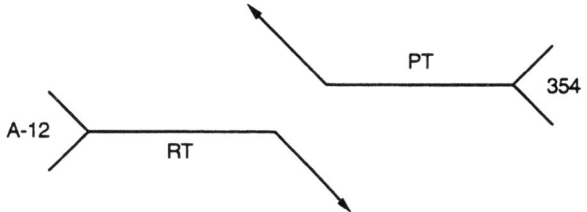

18. Extent, Location, and Orientation, of Nondestructive Examination

18.1 Specifying Length of Section to be Examined

18.1.1 Length Shown. To specify examination of welds or parts where only the length of a section need be considered, the length dimension shall be placed to the right of the letter designation.

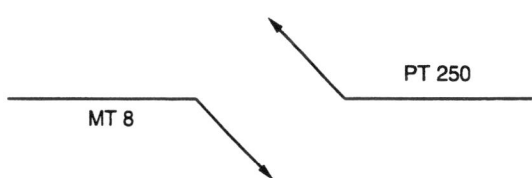

18.1.2 Location Shown. To specify the exact location of a section to be examined, as well as the length, dimension lines shall be used.

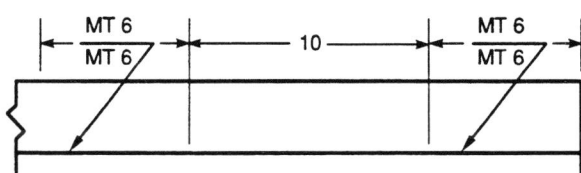

18.1.3 Full Length Examination. When the full length of a part is to be examined, no length dimension need be included in the nondestructive examination symbol.

18.1.4 Partial Examination. When less than one hundred percent of the length of a weld or part is to be examined, with locations to be determined by a specified procedure, the length to be examined is specified by placing the appropriate percentage to the right of the letter designation. The selected procedure may be specified by reference in the tail of the nondestructive examination symbol.

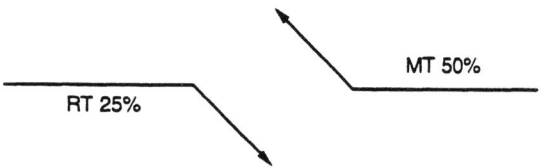

18.2 Number of Examinations. To specify a number of examinations to be conducted on a joint or part at random locations, the number of required examinations shall be placed in parentheses either above or below the letter designation away from the reference line.

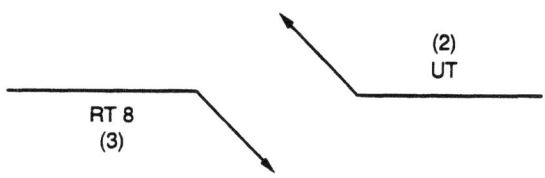

18.3 Examination of Areas. Nondestructive examination of areas shall be specified by one of the following methods:

18.3.1 Plane Areas. To specify nondestructive examination of an area represented as a plane on the drawing, the area to be examined shall be enclosed by straight, broken lines with a circle at each change in direction. The letter designations for the nondestructive examinations required shall be used in connection with these lines as shown below. When necessary, these enclosures shall be located by coordinate dimensions.

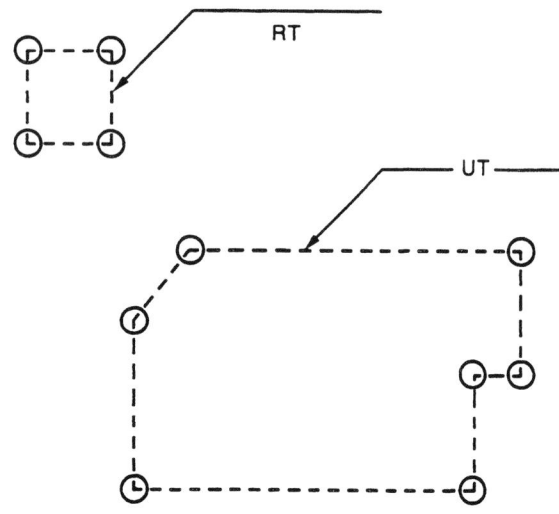

18.3.2 Areas of Revolution. For nondestructive examination of areas of revolution, the area shall be specified by using the examine-all-around symbol and appropriate dimensions. The following illustration specifies:

(1) Magnetic particle examination of the bore of the flange for a distance of two inches from the right-hand face, all the way around the circumference.

(2) Radiographic examination of an area of revolution where dimensions were not available on the drawing.

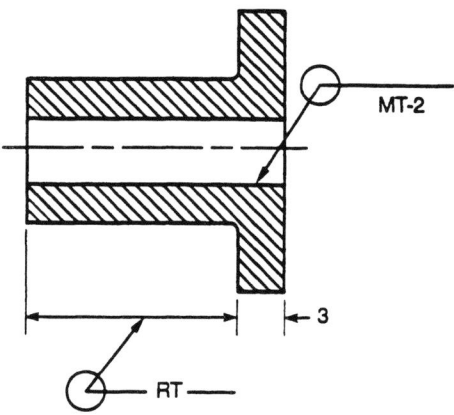

The symbol below specifies an area of revolution subject to an internal proof examination and an external eddy current examination. Since no dimensions are given, the entire length is to be examined.

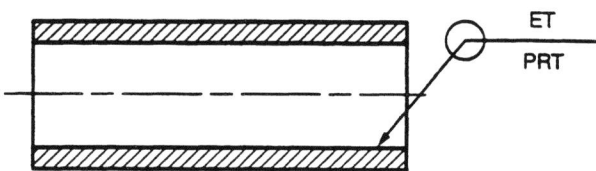

18.3.3 Acoustic Emission. Acoustic emission is generally applied to all or a large portion of a component, such as a pressure vessel or pipe. The symbol below indicates application of AET to the component without specific reference to location of sensors.

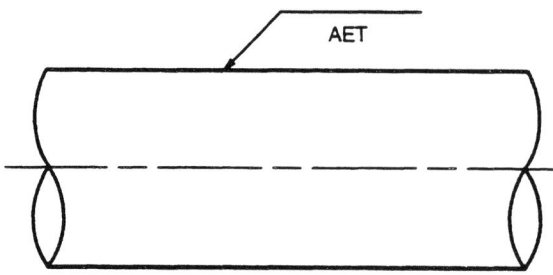

Table 1
Letter Designations of Welding and Allied Processes and Their Variations

Processes and Variations	Letter Designation	Processes and Variations	Letter Designation
adhesive bonding	AB	projection welding	PW
arc welding	AW	resistance seam welding	RSEW
arc stud welding	SW	high frequency seam welding	RSEW-HF
atomic hydrogen welding	AHW	induction seam welding	RSEW-I
bare metal arc welding	BMAW	resistance spot welding	RSW
carbon arc welding	CAW	upset welding	UW
gas carbon arc welding	CAW-G	high frequency upset welding	UW-HF
shielded carbon arc welding	CAW-S	induction upset welding	UW-I
twin carbon arc welding	CAW-T	soldering	S
electrogas welding	EGW	dip soldering	DS
flux cored arc welding	FCAW	furnace soldering	FS
gas shielded flux cored arc welding	FCAW-G	induction soldering	IS
self-shielded flux cored arc welding	FCAW-S	infrared soldering	IRS
gas metal arc welding	GMAW	iron soldering	INS
pulsed gas metal arc welding	GMAW-P	resistance soldering	RS
short circuit gas metal arc welding	GMAW-S	torch soldering	TS
gas tungsten arc welding	GTAW	ultrasonic soldering	USS
pulsed gas tungsten arc welding	GTAW-P	wave soldering	WS
plasma arc welding	PAW	solid-state welding	SSW
shielded metal arc welding	SMAW	coextrusion welding	CEW
submerged arc welding	SAW	cold welding	CW
series submerged arc welding	SAW-S	diffusion welding	DFW
brazing	B	explosion welding	EXW
block brazing	BB	forge welding	FOW
diffusion brazing	DFB	friction welding	FRW
dip brazing	DB	hot pressure welding	HPW
exothermic brazing	EXB	roll welding	ROW
flow brazing	FLB	ultrasonic welding	USW
furnace brazing	FB	thermal cutting	TC
induction brazing	IB	arc cutting	AC
infrared brazing	IRB	air carbon arc cutting	CAC-A
resistance brazing	RB	carbon arc cutting	CAC
torch brazing	TB	gas metal arc cutting	GMAC
twin carbon arc brazing	TCAB	gas tungsten arc cutting	GTAC
braze welding	BW	plasma arc cutting	PAC
arc braze welding	ABW	shielded metal arc cutting	SMAC
carbon arc braze welding	CABW	electron beam cutting	EBC
exothermic braze welding	EXBW	laser beam cutting	LBC
other welding processes		laser beam air cutting	LBC-A
electron beam welding	EBW	laser beam evaporative cutting	LBC-EV
high vacuum electron beam welding	EBW-HV	laser beam inert gas cutting	LBC-IG
medium vacuum electron beam welding	EBW-MV	laser beam oxygen cutting	LBC-O
nonvacuum electron beam welding	EBW-NV	oxygen cutting	OC
electroslag welding	ESW	flux cutting	FOC
flow welding	FLOW	metal powder cutting	POC
induction welding	IW	oxyfuel gas cutting	OFC
laser beam welding	LBW	oxyacetylene cutting	OFC-A
percussion welding	PEW	oxyhydrogen cutting	OFC-H
thermit welding	TW	oxynatural gas cutting	OFC-N
oxyfuel gas welding	OFW	oxypropane cutting	OFC-P
air acetylene welding	AAW	oxygen arc cutting	AOC
oxyacetylene welding	OAW	oxygen lance cutting	LOC
oxyhydrogen welding	OHW	thermal spraying	THSP
pressure gas welding	PGW	arc spraying	ASP
resistance welding	RW	flame spraying	FLSP
flash welding	FW	plasma spraying	PSP

Table 2
Alphabetical Cross Reference to Table 1 by Process

Processes and Variations	Letter Designation	Processes and Variations	Letter Designation
adhesive bonding	AB	iron soldering	INS
arc braze welding	ABW	laser beam air cutting	LBC-A
arc cutting	AC	laser beam cutting	LBC
arc spraying	ASP	laser beam evaporative cutting	LBC-EV
arc stud welding	SW	laser beam inert gas cutting	LBC-IG
arc welding	AW	laser beam oxygen cutting	LBC-O
air acetylene welding	AAW	laser beam welding	LBW
air carbon arc cutting	CAC-A	medium vacuum electron beam welding	EBW-MV
atomic hydrogen welding	AHW	metal powder cutting	POC
bare metal arc welding	BMAW	nonvacuum electron beam welding	EBW-NV
block brazing	BB	oxyacetylene cutting	OFC-A
braze welding	BW	oxyacetylene welding	OAW
brazing	B	oxyfuel gas cutting	OFC
carbon arc braze welding	CABW	oxyfuel gas welding	OFW
carbon arc cutting	CAC	oxygen arc cutting	AOC
carbon arc welding	CAW	oxygen cutting	OC
coextrusion welding	CEW	oxygen lance cutting	LOC
cold welding	CW	oxyhydrogen cutting	OFC-H
diffusion brazing	DFB	oxyhydrogen welding	OHW
diffusion welding	DFW	oxynatural gas cutting	OFC-N
dip brazing	DB	oxypropane cutting	OFC-P
dip soldering	DS	percussion welding	PEW
electrogas welding	EGW	plasma arc cutting	PAC
electron beam cutting	EBC	plasma arc welding	PAW
electron beam welding	EBW	plasma spraying	PSP
electroslag welding	ESW	pressure gas welding	PGW
exothermic braze welding	EXBW	projection welding	PW
exothermic brazing	EXB	pulsed gas metal arc welding	GMAW-P
explosion welding	EXW	pulsed gas tungsten arc welding	GTAW-P
flame spraying	FLSP	resistance brazing	RB
flash welding	FW	resistance seam welding	RSEW
flow brazing	FLB	resistance soldering	RS
flow welding	FLOW	resistance spot welding	RSW
flux cored arc welding	FCAW	resistance welding	RW
flux cutting	FOC	roll welding	ROW
forge welding	FOW	self shielded flux cored arc welding	FCAW-S
friction welding	FRW	series submerged arc welding	SAW-S
furnace brazing	FB	shielded carbon arc welding	CAW-S
furnace soldering	FS	shielded metal arc cutting	SMAC
gas carbon arc welding	CAW-G	shielded metal arc welding	SMAW
gas metal arc cutting	GMAC	short circuit gas metal arc welding	GMAW-S
gas metal arc welding	GMAW	soldering	S
gas shielded flux cored arc welding	FCAW-G	solid-state welding	SSW
gas tungsten arc cutting	GTAC	submerged arc welding	SAW
gas tungsten arc welding	GTAW	thermal cutting	TC
high frequency seam welding	RSEW-HF	thermal spraying	THSP
high frequency upset welding	UW-HF	thermit welding	TW
high vacuum electron beam welding	EBW-HV	torch brazing	TB
hot pressure welding	HPW	torch soldering	TS
induction brazing	IB	twin carbon arc brazing	TCAB
induction seam welding	RSEW-I	twin carbon arc welding	CAW-T
induction soldering	IS	ultrasonic soldering	USS
induction upset welding	UW-I	ultrasonic welding	USW
induction welding	IW	upset welding	UW
infrared brazing	IRB	wave soldering	WS
infrared soldering	IRS		

Table 3
Alphabetical Cross Reference to Table 1 by Letter Designation

Processes and Variations	Letter Designation	Processes and Variations	Letter Designation
AAW	air acetylene welding	INS	iron soldering
AB	adhesive bonding	IRB	infrared brazing
ABW	arc braze welding	IRS	infrared soldering
AC	arc cutting	IS	induction soldering
AHW	atomic hydrogen welding	IW	induction welding
AOC	oxygen arc cutting	LBC	laser beam cutting
ASP	arc spraying	LBC-A	laser beam air cutting
AW	arc welding	LBC-EV	laser beam evaporative cutting
B	brazing	LBC-IG	laser beam inert gas cutting
BB	block brazing	LBC-O	laser beam oxygen cutting
BMAW	bare metal arc welding	LBW	laser beam welding
BW	braze welding	LOC	oxygen lance cutting
CABW	carbon arc braze welding	OAW	oxyacetylene welding
CAC	carbon arc cutting	OC	oxygen cutting
CAC-A	air carbon arc cutting	OFC	oxyfuel gas cutting
CAW	carbon arc welding	OFC-A	oxyacetylene cutting
CAW-G	gas carbon arc welding	OFC-H	oxyhydrogen cutting
CAW-S	shielded carbon arc welding	OFC-N	oxynatural gas cutting
CAW-T	twin carbon arc welding	OFC-P	oxypropane cutting
CEW	coextrusion welding	OFW	oxyfuel gas welding
CW	cold welding	OHW	oxyhydrogen welding
DB	dip brazing	PAC	plasma arc cutting
DFB	diffusion brazing	PAW	plasma arc welding
DFW	diffusion welding	PEW	percussion welding
DS	dip soldering	PGW	pressure gas welding
EBC	electron beam cutting	POC	metal powder cutting
EBW	electron beam welding	PSP	plasma spraying
EBW-HV	high vacuum electron beam welding	PW	projection welding
EBW-MV	medium vacuum electron beam welding	RB	resistance brazing
EBW-NV	nonvacuum electron beam welding	ROW	roll welding
EGW	electrogas welding	RS	resistance soldering
ESW	electroslag welding	RSEW	resistance seam welding
EXB	exothermic brazing	RSEW-HF	high frequency seam welding
EXBW	exothermic braze welding	RSEW-I	induction seam welding
EXW	explosion welding	RSW	resistance spot welding
FB	furnace brazing	RW	resistance welding
FCAW	flux cored arc welding	S	soldering
FCAW-G	gas shielded flux cored arc welding	SAW	submerged arc welding
FCAW-S	self-shielded flux cored arc welding	SAW-S	series submerged arc welding
FLB	flow brazing	SMAC	shielded metal arc cutting
FLOW	flow welding	SMAW	shielded metal arc welding
FLSP	flame spraying	SSW	solid-state welding
FOC	flux cutting	SW	arc stud welding
FOW	forge welding	TB	torch brazing
FRW	friction welding	TC	thermal cutting
FS	furnace soldering	TCAB	twin carbon arc brazing
FW	flash welding	THSP	thermal spraying
GMAC	gas metal arc cutting	TS	torch soldering
GMAW	gas metal arc welding	TW	thermit welding
GMAW-P	pulsed gas metal arc welding	USW	ultrasonic welding
GMAW-S	short circuit gas metal arc welding	UW	upset welding
GTAC	gas tungsten arc cutting	UW-HF	high frequency upset welding
GTAW	gas tungsten arc welding	UW-I	induction upset welding
GTAW-P	pulsed gas tungsten arc welding	WS	wave soldering
HPW	hot pressure welding		
IB	induction brazing		

Table 4
Suffixes for Optional Use in Applying Welding and Allied Processes

Adaptive control	AD	Mechanized	ME
Automatic	AU	Robotic	AU
Manual	MA	Semiautomatic	SA

Table 5
Obsolete or Seldom Used Processes

Processes and Variations	Letter Designations	Processes and Variations	Letter Designations
Air acetylene welding	AAW	Flow brazing	FLB
Atomic hydrogen welding	AHW	Flow welding	FLOW
Bare metal arc welding	BMAW	Twin carbon arc brazing	TCAB
Block brazing	BB	Gas carbon arc welding	CAW-G

Table 6
Joint Type Designators (See Figure 4)

Designator	Joint Type
B	Butt
C	Corner
T	T-joint
L	Lap
E	Edge

Annex A
Design of Standard Symbols (Inches)

(This Annex is not a part of ANSI/AWS A2.4-98, *Standard Symbols for Welding, Brazing, and Nondestructive Examination*, but is included for information purposes only.)

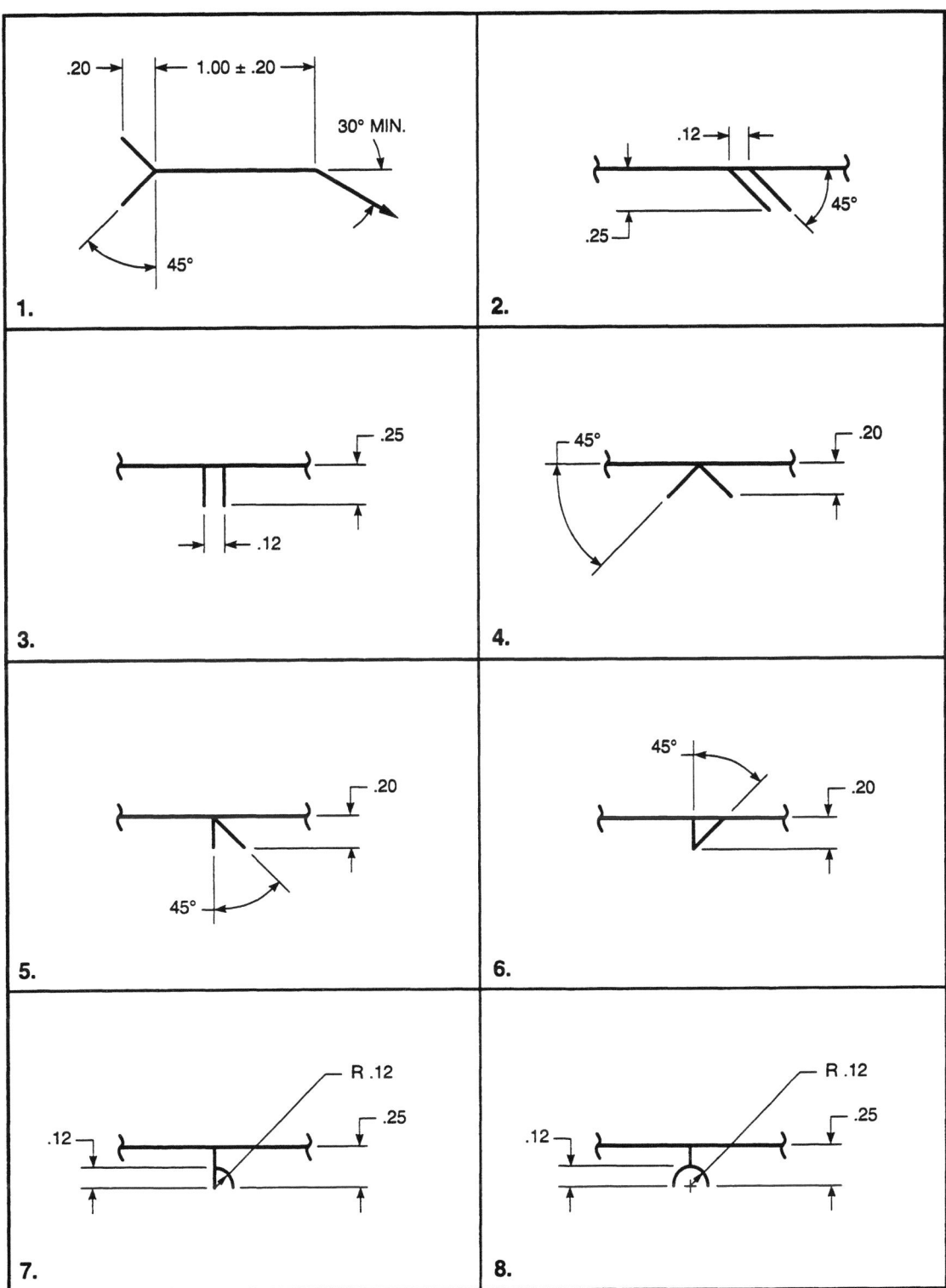

NOTES: 1. UNLESS OTHERWISE SPECIFIED, TOLERANCES SHALL BE ± .04 OR ±1° AS APPLICABLE.
2. ALL RADII ARE MINIMUM DIMENSIONS.

Annex A (Continued)
Design of Standard Symbols (Inches)

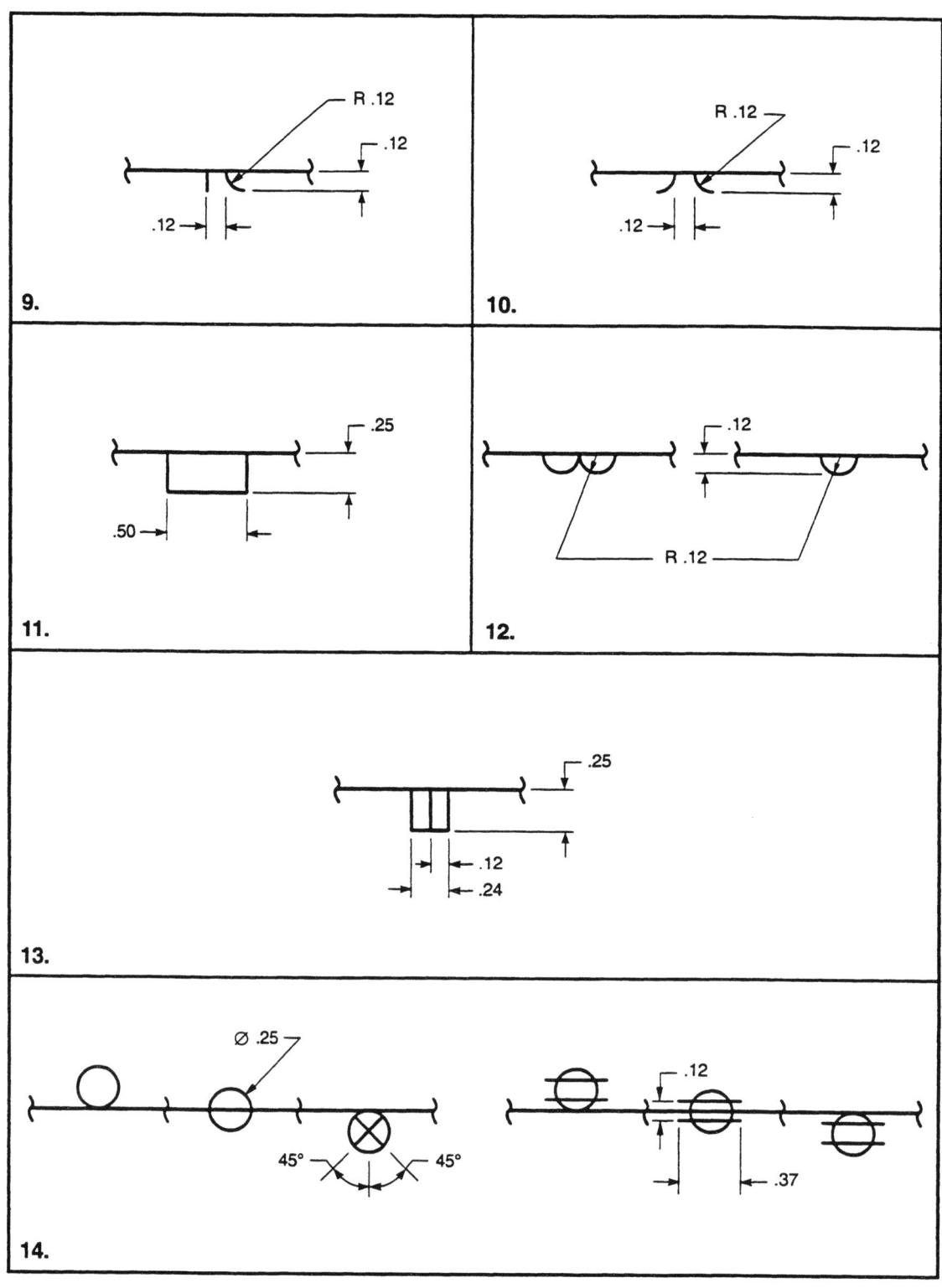

NOTES: 1. UNLESS OTHERWISE SPECIFIED, TOLERANCES SHALL BE ± .04 OR ±1° AS APPLICABLE.
2. ALL RADII ARE MINIMUM DIMENSIONS.

Annex A (Continued)
Design of Standard Symbols (Inches)

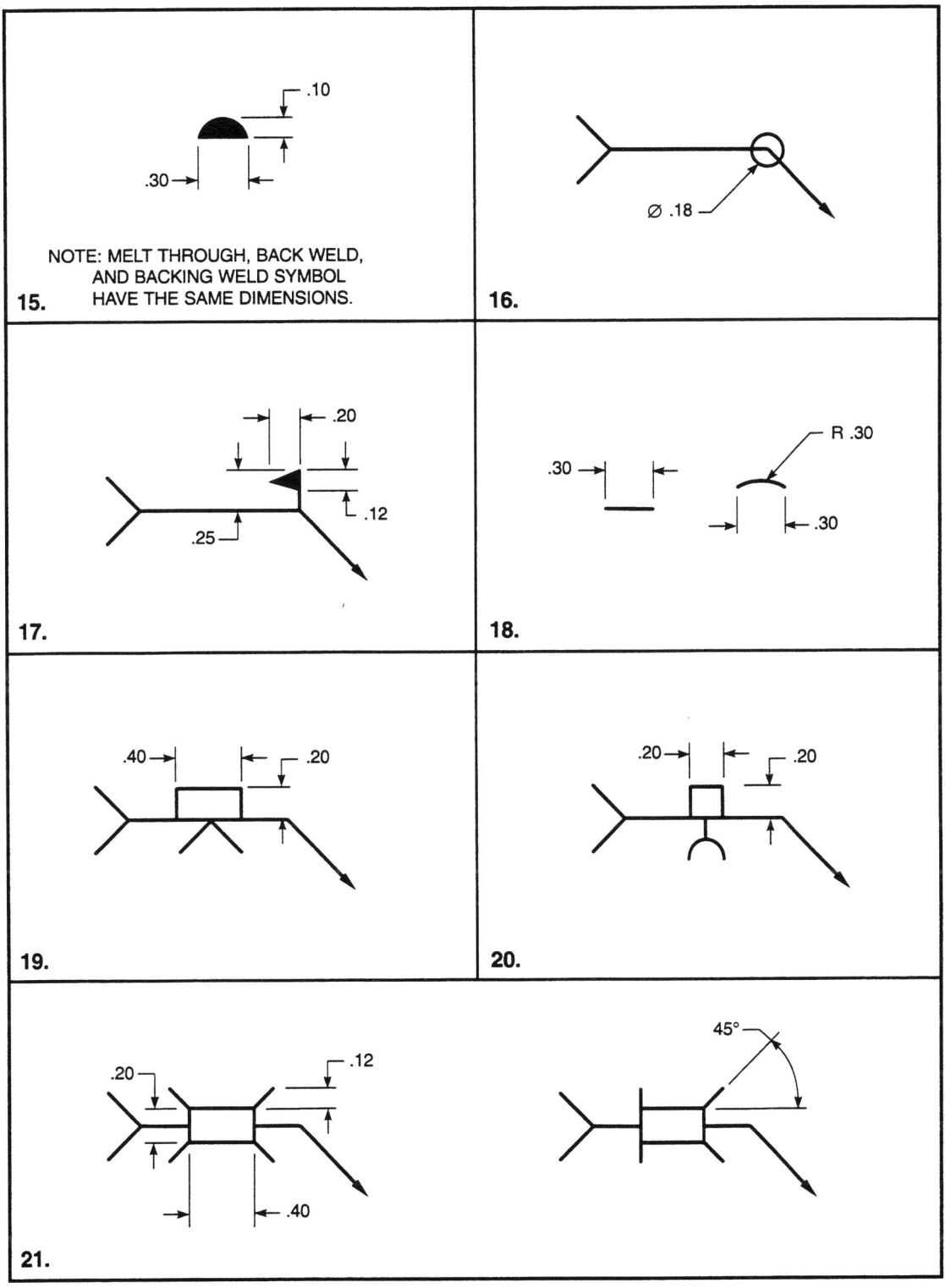

NOTES:
1. UNLESS OTHERWISE SPECIFIED, TOLERANCES SHALL BE ± .04 OR ±1° AS APPLICABLE.
2. ALL RADII ARE MINIMUM DIMENSIONS.

Annex AM
Design of Standard Symbols (Millimeters)

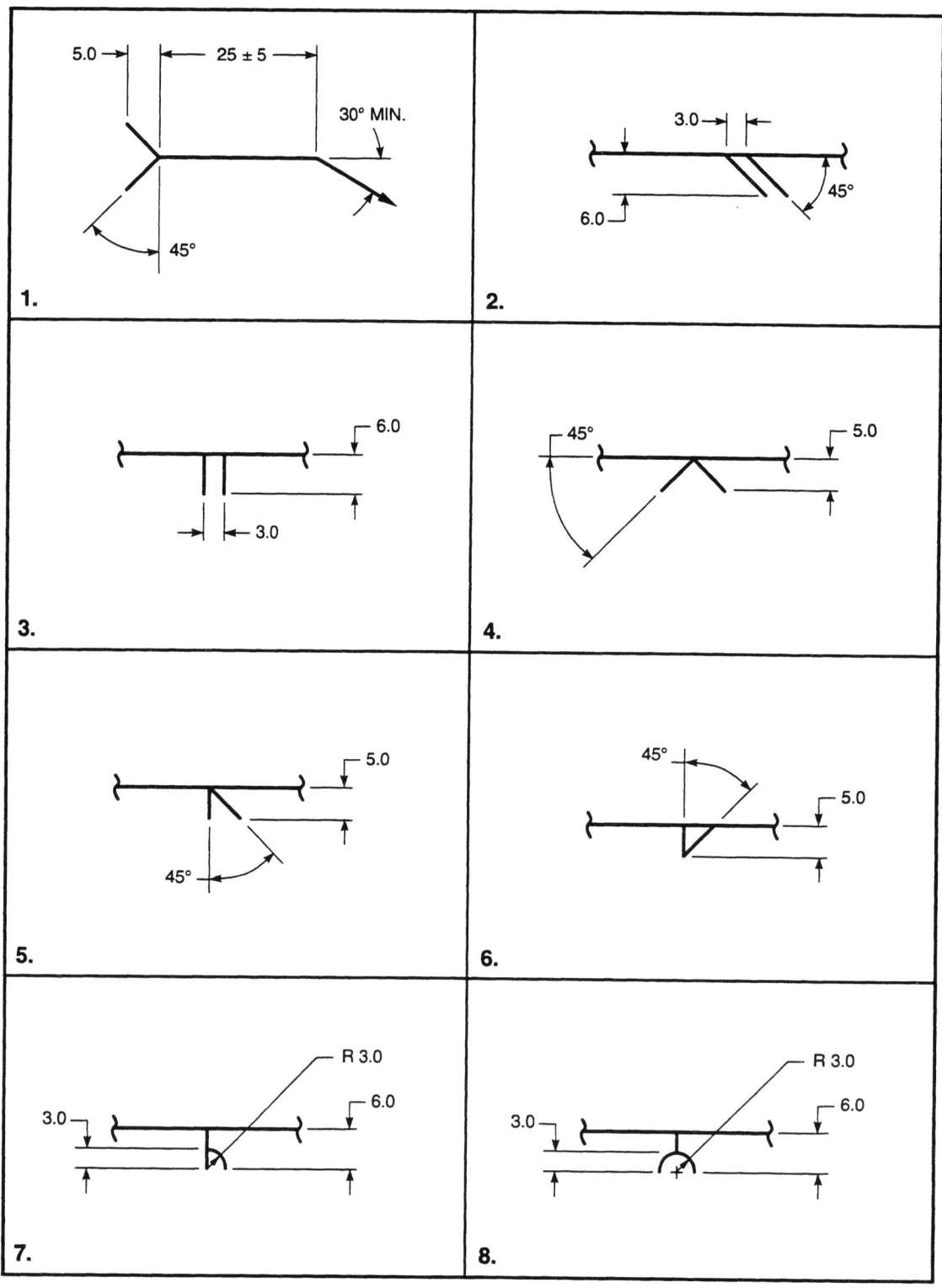

NOTES: 1. UNLESS OTHERWISE SPECIFIED, TOLERANCES SHALL BE ± 1 mm OR ±1° AS APPLICABLE.
2. ALL RADII ARE MINIMUM DIMENSIONS.

**Annex AM (Continued)
Design of Standard Symbols (Millimeters)**

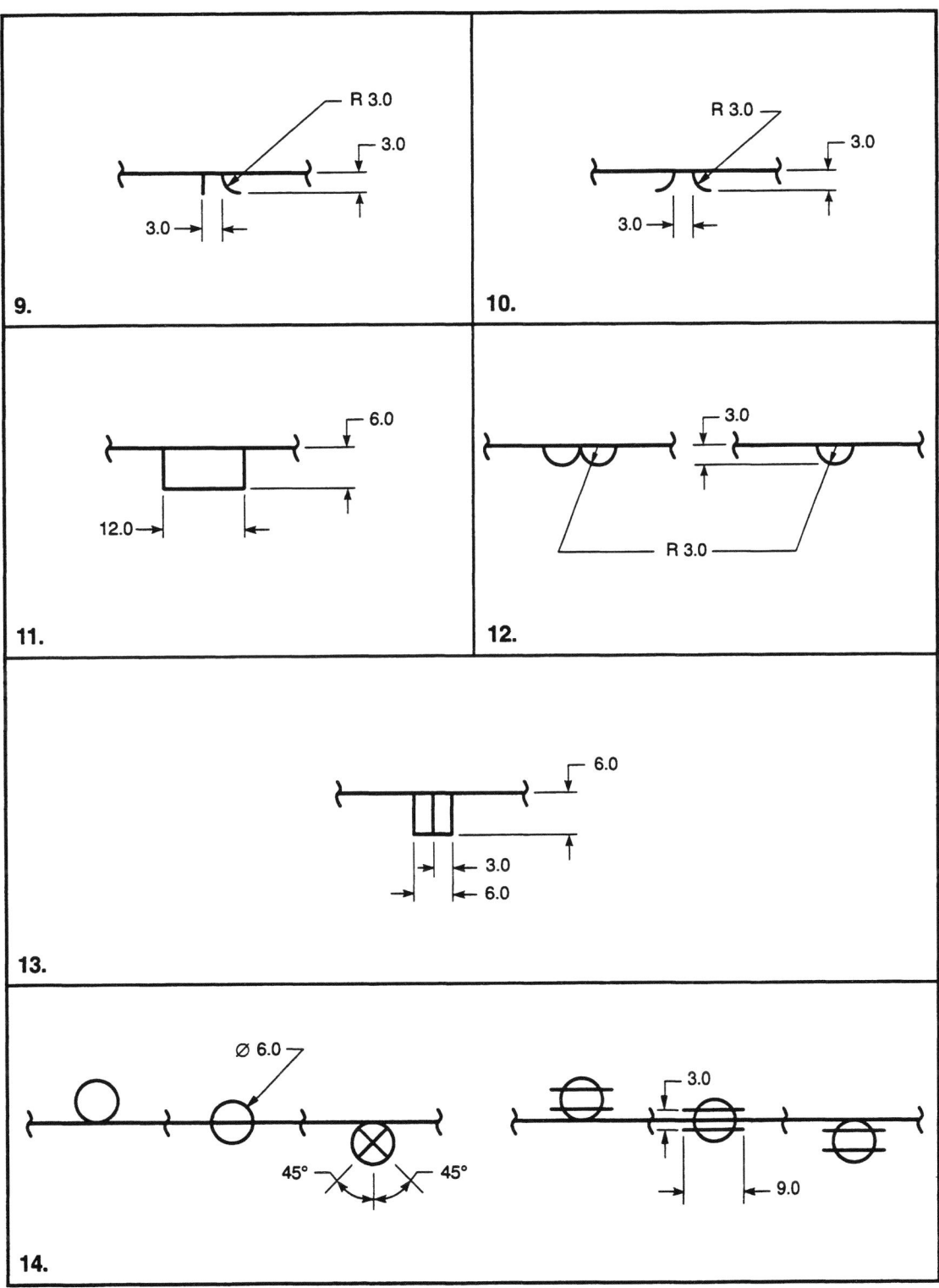

NOTES: 1. UNLESS OTHERWISE SPECIFIED, TOLERANCES SHALL BE ± 1 mm OR ±1° AS APPLICABLE.
2. ALL RADII ARE MINIMUM DIMENSIONS.

Annex AM (Continued)
Design of Standard Symbols (Millimeters)

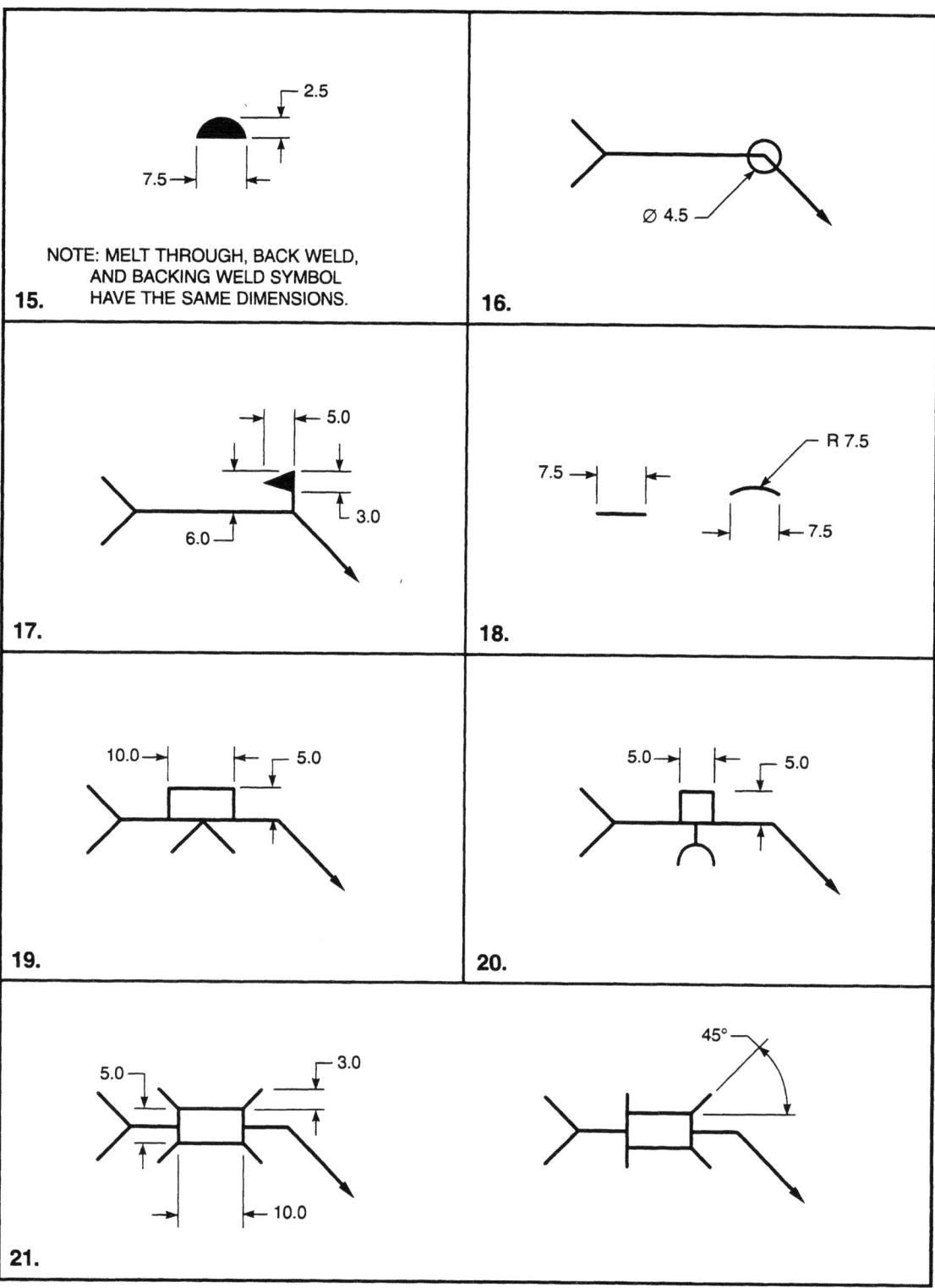

NOTES: 1. UNLESS OTHERWISE SPECIFIED, TOLERANCES SHALL BE ± 1 mm OR ±1° AS APPLICABLE.
2. ALL RADII ARE MINIMUM DIMENSIONS.

Annex B
Commentary on A2.4-98, Standard Symbols for Welding, Brazing, and Nondestructive Examination

(This Annex is not a part of ANSI/AWS A2.4-97, *Standard Symbols for Welding, Brazing, and Nondestructive Examination*, but is included for information purposes only.)

Note: Numbered paragraphs in this Annex relate to similarly numbered paragraphs in the text of A2.4, e.g., paragraph B3.8 is a commentary on paragraph 3.8 in the text.

B3.8 Field Weld Symbols. Welds are designated to be made in the field by the addition of the field weld symbol when the welding symbols are added to a drawing. It should be understood, however; that the placing of field weld symbols on drawings at the design stage does not preclude further discussion by the parties involved, and possibly different decisions regarding where the welding will be done. If changes are made, the drawings should be revised and the field weld symbols added or deleted as appropriate.

B3.9.2 Square and Rectangular Tubing. The use of square and rectangular tubing has resulted in numerous applications involving joints in which the axes of the tubes are perpendicular as a branch-to-header or a T connection. The tubes are often of equal size, as illustrated below, and it is intended that welds extend around the outside surface of the branch tube or stem of the T. The welds are usually fillet or square-groove on two of the opposite sides and flare-bevel-groove on the other two opposite sides. The weld-all-around symbol is not appropriate to specify the welds described since the joints are not all of the same type and the welds may differ in size. Instead two welding symbols should be used, each with two arrows pointing to the specific joints intended, one to specify the fillet or square-groove welds and the second to specify the flare-bevel-groove welds as shown in the following illustration.

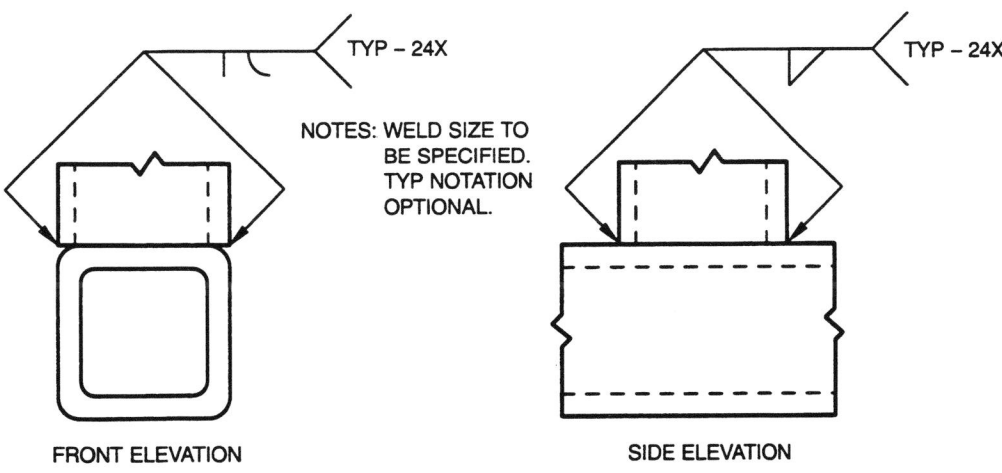

B3.10.1 Weld-All-Around Symbol. A continuous weld is one which has no breaks in its length and does not change in size, geometry, or weld type. Such a weld, which extends around a series of connected joints and ends at the point of origin, may be specified by adding the weld-all-around symbol. The joint may require welding in different directions and positions and the welds may lie in more than one plane. The most common applications involve either fillet welds or square groove welds and are often intended to provide a gas or liquid seal in addition to or in lieu of carrying the loads imposed on the joints. The weld-all-around symbol should not be used in place of double-fillet or symmetrical double-groove weld symbols for specifying welds on both sides of the same thickness of base metal.

B3.11.3 Welding Symbols Designated "Typical". The "TYPICAL" designation is intended as an alternative to repeating identical welding symbols many times on the same drawing, but only when the joints represented are identical in all details. The "TYPICAL" notation is added to the tail of the welding symbol, usually abbreviated "TYP", and all applicable joints must be completely identified, i.e., "TYP at four stiffeners".

Misuse of the "TYPICAL" designation has caused many instances of confusion and fabrication errors by failing to completely identify all applicable joints or by identifying joints that might be similar but not identical. If more extensive information is required it may be stated in a separate drawing note with a reference in the tail of the welding symbol.

B3.19 Changes in Joint Geometry During Welding. Joint geometry of groove welds is sometimes changed as a result of specified welding operations. These changes in joint geometry are not to be included in the welding symbol. For example; a welding symbol could specify a V-groove weld on the arrow side of a joint and a square-groove weld on the other side of the joint with backgouging to sound metal, from the other side of the joint, using air carbon arc cutting. With the V-groove weld completed, the backgouging operation would be expected to produce a weld groove that could be described as a U-groove. This change in geometry, from a square-groove to a U-groove, is not to be specified in the welding symbol (see following illustration).

B4.2.2 Complete Joint Penetration. Complete joint penetration is defined as, "Penetration of weld metal through the thickness of a joint with a groove weld". The simplest way of specifying such a groove weld is to show no dimensions to the left of the groove weld symbol. This is the intent of 4.2.2. There are other ways by which complete joint penetration can be specified including:

WELDING SYMBOL:

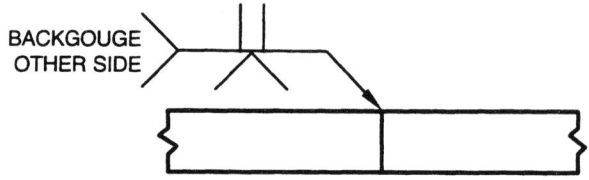

BACKGOUGE OTHER SIDE

BEFORE WELDING:

V-GROOVE WELD COMPLETE:

BACKGOUGING COMPLETE:

WELD COMPLETE:

4.2.4—Nonsymmetrical double-groove welds
4.2.8—"CJP" in the tail of the welding symbol
4.7 —Back or backing welds
4.8 —Joint with backing

The provision in 4.2.8 is included for use on design drawings where there is insufficient information available as to what equipment might be used or, in some cases, what company or organization might do the work. For example, the design drawings might be completed prior to the job being submitted for bids. In these situations, it is considered good practice to require the successful bidder to submit construction drawings complete with detailed welding symbols for review. The other methods identified above require knowledge of the specific welding situation and also the requirements of any codes or specifications that might apply.

B4.2.10 Flare-Groove Welds. Although flare-groove welds are included in the section on groove welds, they must be treated as special cases since they do not conform to all of the accepted conventions associated with other types of groove welds. The dimensions corresponding to "depth of bevel" and "groove angle" in a normal groove weld are functions of the curvature of the base metal in a flare-groove weld and therefore beyond the usual controls of either the designer or the welder. Of even greater importance is the concept of complete joint penetration which is not attainable in many flare-groove welds since the fusion occurs along the surface of one or both members rather than through the thickness. The rate of curvature on one or both members is such that the actual obtainable weld size is usually only some fraction of the radius.

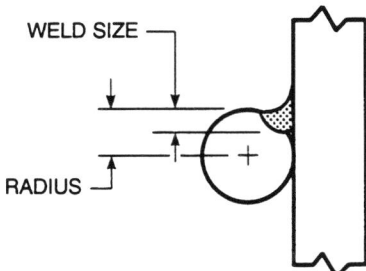

B4.12 Seal Welds. The primary function of a weld may be to contain fluids or gases, however; it will not perform this function if it cracks as a result of stresses caused by handling, storage, shipping, vibrations, temperature changes, etc. For these reasons a seal weld may require careful consideration regarding dimensions of the groove as well as the type. It should be recognized that a welding symbol with only "SEAL WELD" in the tail and no other requirements will relegate such welding to the discretion of the fabrication shop, whose judgement and welding practice may not ensure the service performance of the joint as expected by the designer.

Commentary on Welding Symbol Chart

The welding symbol chart included in ANSI/AWS A2.4 is intended to provide basic information and often used symbols in a convenient form as a shop or drafting room aid. The chart is published separately from, but concurrently with, ANSI/AWS A2.4 in both wall size and desk size formats. Over the years, the charts have been reproduced and distributed by other sources both with and without AWS permission. Consequently there are many obsolete and error filled versions in existence. The reader is advised and cautioned that the only complete and approved version is in the latest edition of ANSI/AWS A2.4.

AMERICAN WELDING SOCIETY
Welding Symbol Chart

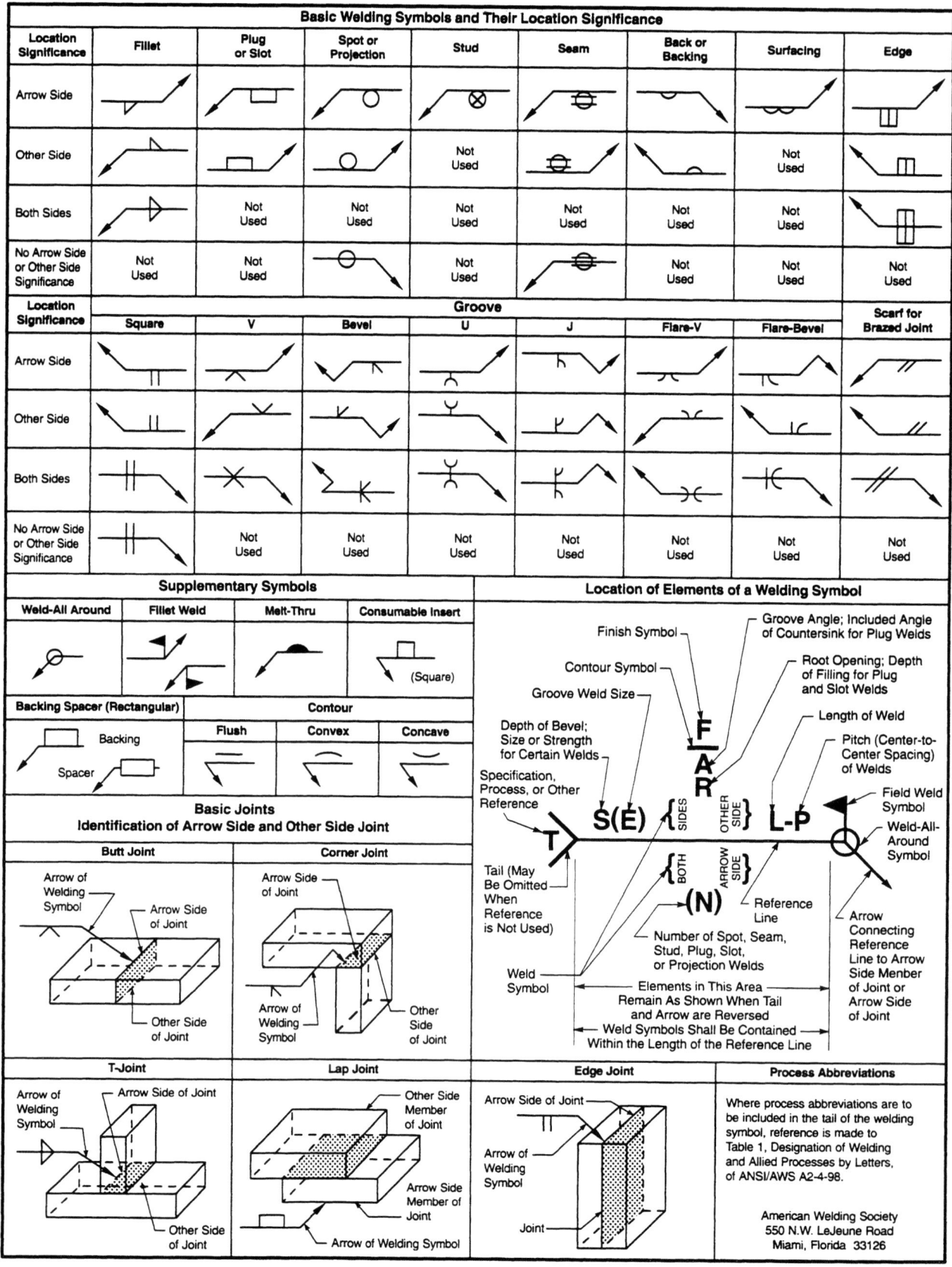

AMERICAN WELDING SOCIETY
Welding Symbol Chart

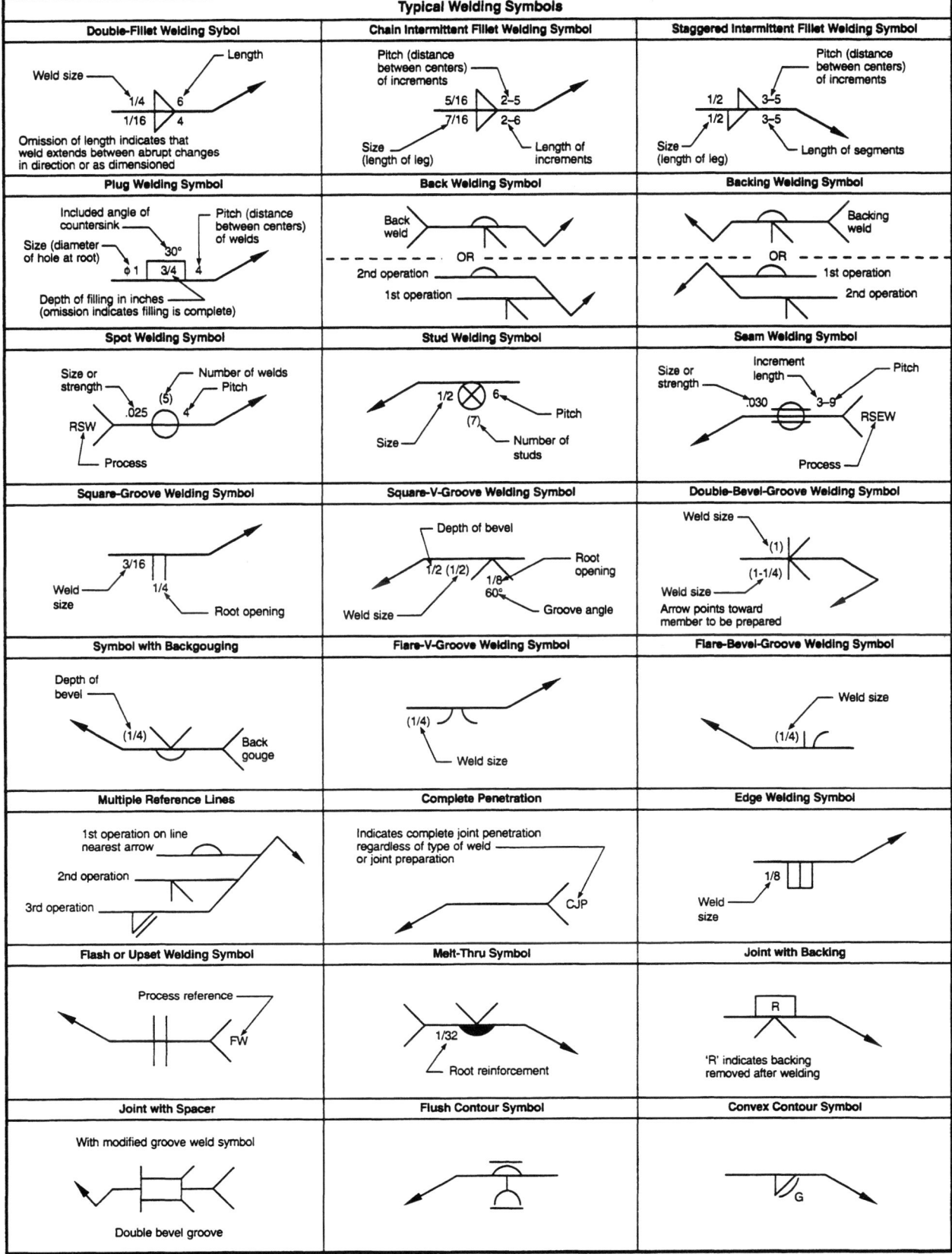

*It should be understood that these charts are intended only as shop aids. The only complete and official presentation of the standard welding symbols is in A2.4.

Definitions and Symbols Document List

Designation	Title
AWS A2.1-WC*	Welding Symbol Chart*. (Wall size)
AWS A2.1-DC*	Welding Symbol Chart*. (Desk size)
ANSI/AWS A2.4	Standard Symbols for Welding, Brazing, and Nondestructive Examination
ANSI/AWS A3.0	Standard Welding Terms and Definitions including Terms for Brazing, Soldering, Thermal Spraying, and Thermal Cutting

*A reproduction of the charts is shown on the final two pages. It should be understood that these charts are intended only as shop aids. The only complete and official presentation of the Standard Welding Symbols is in A2.4.